International Series in Operations Research & Management Science

Volume 317

More information about this series at https://link.springer.com/bookseries/6161

Alireza Amirteimoori • Biresh K. Sahoo •
Vincent Charles • Saber Mehdizadeh

Stochastic Benchmarking

Theory and Applications

Springer

Alireza Amirteimoori
Rasht Branch, Islamic Azad University
Rasht, Iran

Biresh K. Sahoo
Xavier Institute of Management
XIM University
Bhubaneswar, India

Vincent Charles
Center for Value Chain Innovation,
CENTRUM Catolica Graduate
Business School
Pontifical Catholic University of Peru
Lima, Peru

Saber Mehdizadeh
Rasht Branch, Islamic Azad University
Rasht, Iran

ISSN 0884-8289 ISSN 2214-7934 (electronic)
International Series in Operations Research & Management Science
ISBN 978-3-030-89868-7 ISBN 978-3-030-89869-4 (eBook)
https://doi.org/10.1007/978-3-030-89869-4

This Springer imprint is published by the registered company Springer Nature Switzerland AG.
The registered company address is: Gewerbestrasse 11, 6330 Cham, Switzerland

We dedicate this book to Professor Rajiv D. Banker, a pioneer in the field of data envelopment analysis. His passion for research has helped to shape our understanding and thinking around efficiency and productivity measurement. A Harvard University graduate Professor Banker has dedicated his career to advancing the field and spreading the knowledge across the Globe while encouraging and supporting researchers along the way. His extraordinary contributions to literature and society are inspirational and enduring and are matched only by his humble and kind personality.

Preface

In the last three decades, benchmarking as a performance measurement procedure has become a powerful quantitative and analytical tool for measuring the relative efficiency of entities. This book introduces benchmarking techniques under a stochastic environment, most notably, the Stochastic Data Envelopment Analysis (SDEA). Similar to different stochastic theories, the SDEA approach deals with real-life problems that involve stochastic inputs and outputs. The SDEA models shown in this book account for the possibility of variations in both inputs and outputs. To enhance understanding, simple examples along with graphical illustrations are also provided. The uses of theories and interpretations of the mathematical programs are emphasized. Moreover, the proposed mathematical programs are combined with economic and organizational thinking. Real industry-relevant applications are further discussed. The underlying theories are introduced, which support relevant calculations and help in the discussion of applications. Overall, our purpose is to inform the reader about the advantages of the different methods under deterministic and stochastic environments, as well as about the proper use of these methods in different cases. This book can be used as a reference text or as a textbook. As a textbook, it covers materials that we have taught over the years in several courses and modules.

There are several excellent books on performance analysis in the literature. Our book is not designed to compete with the existing books, but rather to complement them by contributing to the topic of performance measurement using SDEA.

We hope that the readers find this book useful in their studies and research. We welcome any critical feedback and constructive comments.

Rasht, Iran — Alireza Amirteimoori
Bhubaneswar, India — Biresh K. Sahoo
Lima, Peru — Vincent Charles
Rasht, Iran — Saber Mehdizadeh

Introduction

In the last three decades, data envelopment analysis (DEA) as the main benchmarking technique has proven to be an excellent and powerful quantitative tool for evaluating and measuring the relative efficiency of entities. Classic DEA models assume that there are no variations in the values of the inputs and outputs. In the presence of stochastic variations in the data, the concepts of efficiency and returns-to-scale (RTS) are inextricably related to how entities deal with uncertainty. Therefore, when there are variations in inputs and outputs due to uncertainty, the evaluation of efficiency and RTS under a deterministic DEA setting becomes sensitive to such variations. As a result, one would expect to see whether the efficiency and RTS characterizations of entities are subject to change under this stochastic environment.

Benchmarking

Benchmarking is an efficiency measurement procedure that allows entities to compare their performance to the top competitors. In Chap. 1, benchmarking (as a performance measurement tool using a specific indicator) and key performance indicators (KPI) are first briefly introduced. Then, we introduce and connect the concepts of efficiency and productivity and discuss the ways for measuring different types of efficiencies.

An Introduction to DEA

DEA, as a powerful estimator in benchmarking, has proven to be an excellent tool in measuring the relative efficiency of homogeneous entities. An introduction to classic DEA models along with a comprehensive review and discussion is given in Chap. 2.

Probability Theory

Probability theory is concerned with the analysis of random phenomena. It is the branch of mathematics concerned with probability. Chapter 3 introduces elementary terms in probability theory in such a way that readers who are beginners to this subject can quickly become familiar with various terminologies.

Stochastic DEA

Stochasticity and stochastic variations in inputs and outputs represent a variety of uncertainty. In this sense, incorporating the stochastic variations of the input/output data into DEA models is an important subject, which has received considerable attention over the past two decades. The underlying theories of DEA under a stochastic environment are presented in Chap. 4. This permits us to replace terms such as "efficient" and "inefficient" with "probably efficient" and "probably inefficient," respectively.

Stochastic Network DEA

Two-stage network structures in DEA are studied in cases in which the input/output data set is deterministic. In many real applications, however, we face uncertainty. Chapter 5 proposes a two-stage network DEA model for input/output data that are stochastic.

Stochastic Scale Elasticity

To analyze the performance of entities, several economic concepts such as "economies of scale" (RTS) and "marginal rates of technical substitutions" have been used in the literature. In the last chapter of this book, we concentrate on determining and measuring the RTS in quantitative form (called scale elasticity). As such, Chap. 6 provides a detailed study of stochastic scale elasticity in DEA.

Contents

Chapter 1
Benchmarking

Benchmarking is an efficiency performance measurement procedure that allows firms to compare their performance to the top competitors. In this chapter, benchmarking (as an efficiency performance measurement tool using a specific indicator) and key performance indicators are first briefly introduced. Then, we introduce efficiency and productivity and the ways for measuring the different types of efficiencies.

1.1 What Is Benchmarking?

Benchmarking is a procedure for identifying best practices to learn from and adapt outstanding practices to improve efficiency performance. In economics terminology, benchmarking is defined through a function that maps the firm (represented by its observed inputs and outputs) and the industry into the efficiency of the firm. This mapping summarizes a program that identifies the best practices or benchmarks for the firm. The efficiency performance of the firm will be measured by the ratio of its weighted outputs to weighted inputs. This weighted output–input ratio is also called total factor productivity, a notion which is very much linked to efficiency when input and output weights are interpreted as their respective accounting prices or opportunity costs in a competitive market.

In relative performance evaluation, if we know a utility function and the technology set, we can easily evaluate the relative performances of the firms. Figure 1.1 illustrates the relative performance evaluation. Two inputs, say, labor and composite input (*i.e.*, cost on other inputs), are used to produce a single-output. The input isoquant shows various minimal combinations of these two inputs (cost and labor) that are required to produce a given level of output. The utility function is given by a linear curve passing through point P. This line shows the optimal inputs

A. Amirteimoori et al., *Stochastic Benchmarking*, International Series in Operations Research & Management Science 317,
https://doi.org/10.1007/978-3-030-89869-4_1

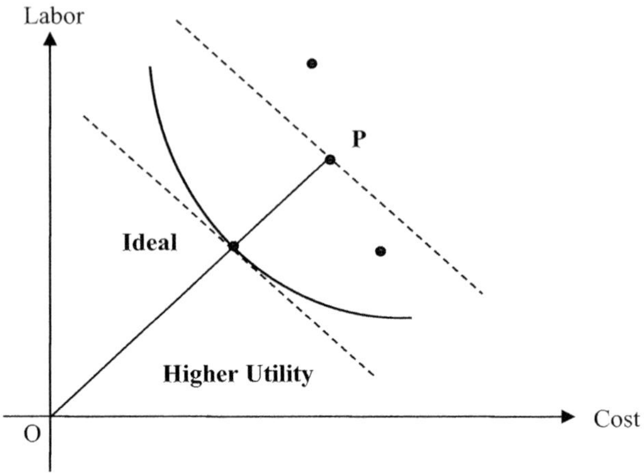

Fig. 1.1 The input isoquant

combination. The relative performance evaluation compares the observed utility level with the optimal utility level. For point *P*, *U*(*P*) is compared to *U*(*Ideal*), and in this sense,

$$Effectiveness = \frac{\underset{Z \in T}{\text{Max}}\{U(Z)\}}{U(P)} = \frac{U(Ideal)}{U(p)}.$$

The term *effectiveness* comes from the fact that we count on both the technology set and the utility function. However, in most real-life applications, the information on these two fundamental requirements is not available, in which case, we rely on some proxy tools. Benchmarking is one of these proxy tools. It attempts to approximate the economics-based idea of relative performance evaluation. The observed input–output data are used to construct an empirical production possibility set (technology set) and the key tool "effectiveness" is replaced by "efficiency." Benchmarking uses frontier analysis methods to construct the technology set and its frontier. Two fundamental approaches are used to construct the production function: parametric and nonparametric approaches. In the parametric approach, a functional form of the production function with a finite number of unknown parameters is considered and the parameters are estimated from the data. In the nonparametric approach, however, we do not need a functional form of the production function, and data are used to construct, based on some production axioms, an empirical production function instead. Another relevant question regarding benchmarking evaluation is whether to use the deterministic or the stochastic approach. A well-known deterministic nonparametric approach in benchmarking evaluation is data envelopment analysis (DEA), which was first introduced by Charnes et al. (1978) under the constant returns-to-scale (CRS) specification of technology and was later extended by Banker et al. (1984) under the variable returns-to-scale (VRS) specification of technology.

1.2 Key Performance Indicator

To make a rational performance evaluation of the firms, we need a key performance indicator (KPI). Decision-makers in industrial and service sectors are generally interested in using KPIs in their performance evaluation. A KPI is a measurable value that illustrates how effectively a firm is performing toward achieving its targets. KPIs are essentially linked to the firm's goals, and managers use these indicators to assess whether they are on target as they work toward the goals. In defining a KPI, managers make a rational decision to determine the cost of input and output measures and their relative importance. A good KPI must be well-defined and quantifiable. The KPI that we use in this book is expressed as

$$\text{KPI} = \frac{\text{The weighted sum of outputs}}{\text{The weighted sum of inputs}}.$$

This expression of the KPI is well-defined and is a simple extension of the single-input-single-output case. When we use this KPI, we implicitly assume that we can scale the inputs and outputs up or down linearly. In other words, we assume the technology to be operating under constant returns-to-scale. However, this assumption can easily be relaxed. One issue with this definition of the KPI is the selection of the weighting scheme. Different weights lead to different KPI measures and this multiplicity may be problematic. We will show later how to overcome this problem.

1.3 Efficiency and Productivity

Efficiency and productivity are two important goals for any firm and the aim is to maximize both. In case of single-input-single-output technology, while a firm's productivity is defined as the ratio of the output produced to the input used, its efficiency is defined as the ratio of its actual productivity to the potential (maximum) productivity in the sample. In case of a technology exhibiting CRS, this definition of efficiency can be equivalently expressed in two alternative ways: (1) the ratio of the output produced to the potential output that would have been produced, had the actual (observed) input been used most efficiently (output technical efficiency, (O)TE) and (2) the ratio of the minimal input required to produce the current (observed) output to the actual input used (input technical efficiency, (I)TE).

Let us illustrate the above case with the help of an example of a simple one-input-one-output technology characterized by the production function: $y = f(x) = 2x$, which is exhibited in Fig. 1.2. This simple technology consists of four firms labeled as A, B, C, and D among which D exhibits the maximum productivity of 2 ($\theta_B = 4/2$) in the sample and is, hence, efficient as its actual and potential productivities are the same. Let us now consider computing the productivity and efficiency scores of firm, say C. The productivity of C is 0.8 ($\theta_C = 2/2.5$). C's technical efficiency is defined as

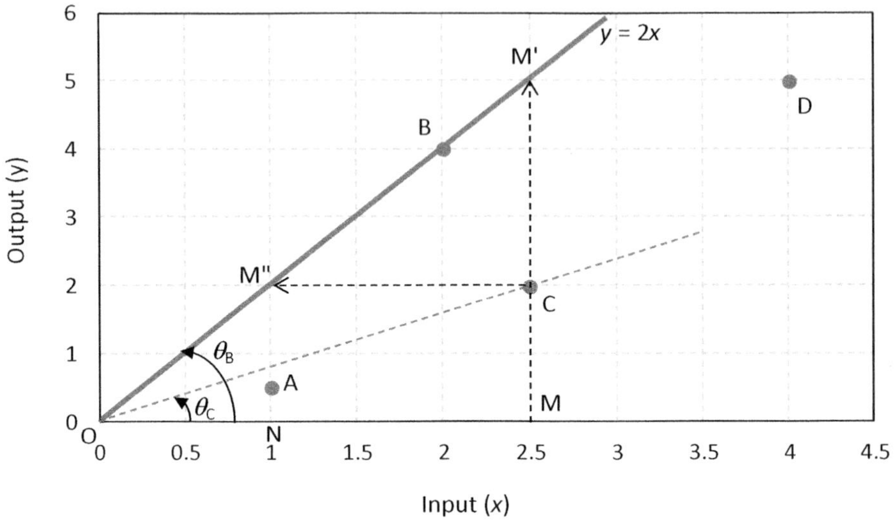

Fig. 1.2 Production function: $q = 2x$ exhibiting CRS

the ratio of its actual productivity to the potential productivity in the sample, *i.e.* $\theta_C/\theta_B = 0.8/2 = 0.4$. We now need to show that C's (I)TE and (O)TE scores are also same as 0.4. (O)TE $= \frac{2}{5} = 0.4$ (where actual output is 2 and potential output is 5) and (I)TE $= \frac{1}{2.5} = 0.4$ (where minimal input is 1 and actual input is 2.5). It is thus clear that in case of technology exhibiting CRS, the input and output technical efficiency scores of a firm are same, and the relationship between efficiency and productivity is direct, *i.e.* the higher the productivity, the higher the efficiency, and vice versa.

However, in case of a technology exhibiting VRS, the input and output technical efficiencies of a firm need not be the same, and the resulting relationship between efficiency and productivity is a bit complex. Let us now illustrate this with an example of a one-input-one-output VRS technology consisting of four firms—P, Q, R, and S. This VRS technology is exhibited in Fig. 1.2. The curve passing through *O*, *Q*, and *S* represents the production frontier. Firms *S* and *Q* are located on the production frontier and are, hence, technically efficient. However, firms *P* and *R* are technically inefficient, because they have the potential to either increase the output or reduce the input or both. Let us now compute the output and input technical efficiency scores of an inefficient firm, P. One can clearly see from Fig. 1.3 that the output and input efficiency scores are 0.60 (=3/5) and 0.727 (=4/5.5), respectively. It is thus clear that the input and output efficiency scores are different. And as regards the relationship between efficiency and productivity, it is also not clear as one can clearly see that though both Q and S are fully efficient, they are not equally productive. Further, a firm, say Q, may be technically efficient but exhibit lower productivity as compared to an inefficient firm R.

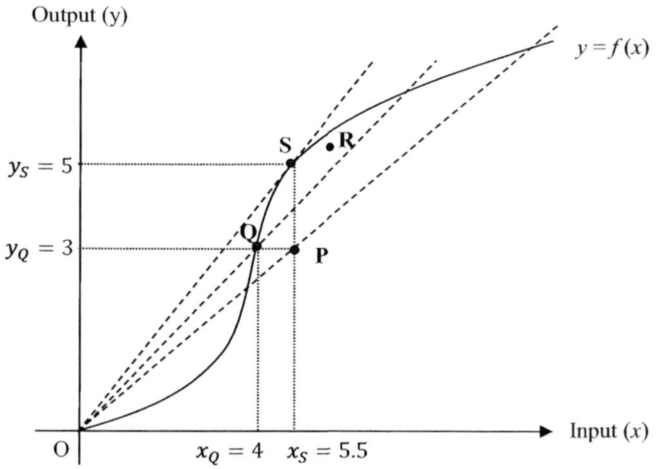

Fig. 1.3 Production function: $y = 2x$ exhibiting VRS

Note that since S exhibits the maximum level of productivity, it is called technical optimal scale (TOS) in the economics literature and the most productive scale size (MPSS) in the OR/MS literature.

We now turn to linking productivity to efficiency under two conditions: (1) the technology exhibits CRS and (2) the inputs and outputs are evaluated at their respective accounting prices or opportunity costs. To demonstrate this, let us consider a two-input-one-output technology: $y^* \equiv f(x_1, x_2) = Ax_1^{\alpha}x_2^{\beta}$. Assuming the unit prices of y, x_1, and x_2 to be p, w_1, and w_2, respectively, the total factor productivity (TFP) of a firm is defined as the ratio of the weighted sum of outputs to the weighted sum of inputs, *i.e.*

$$TFP = \frac{py}{w_1x_1 + w_2x_2} = \frac{y}{\left(\frac{w_1}{p}\right)x_1 + \left(\frac{w_1}{p}\right)x_2}.$$

If the market is perfectly competitive, and the firms are assumed to be profit maximizers, then the marginal productivity (MP) of the inputs can be, respectively, expressed as

$$MP_1 = \frac{w_1}{p} \text{ and } MP_2 = \frac{w_2}{p}.$$

On substitution of these in TFP yields

$$TFP = \frac{y}{MP_1x_1 + MP_2x_2}.$$

Since $f(x_1, x_2)$ is assumed to be homogenous of degree $(\alpha + \beta)$, the *TFP* expression can be written as

$$TFP = \frac{y}{(\alpha + \beta)y^*}.$$

If CRS are assumed, $\alpha + \beta = 1$. Then, *TFP* can be interpreted as *TE*, *i.e.*

$$TFP = \frac{y}{y^*} = \frac{\text{Actual output}}{\text{Potential output}} = TE.$$

Besides the technical efficiency measure, there are other efficiency measures—cost efficiency and revenue efficiency measures. While cost-efficient firms choose an optimal combination of resources, revenue-efficient firms aim at achieving an optimal combination of products.

1.4 Efficiency Analysis

From an economic perspective, cost (revenue) efficiency is due to two sources: input (output) technical efficiency and input (output) allocative efficiency. Technical efficiency refers to the firm's success in producing more outputs by consuming fewer inputs. In this sense, technical efficiency is focused on input orientation or output orientation. If we were to have more information on the data set, such as the prices of the inputs and/or outputs, we could think of computing allocative efficiency, which refers to the ability of a firm to consume inputs and generate outputs in an optimal proportion.

In what follows, these two types of efficiencies are studied in greater depth.

1.4.1 Technical Efficiency

As stated before, technical efficiency refers to the effectiveness with which a firm produces the maximum output from the minimum quantity of inputs.

Definition 1.1. Technical efficiency (Koopmans, 1951): A firm is said to be technically efficient if an increase in any outputs requires a decrease in some outputs or an increase in some inputs.

This definition states that a technically inefficient firm can produce its current outputs with fewer inputs, or it can produce more outputs by using its current inputs.

Definition 1.2. Technical efficiency (Farrell, 1957):

- In input orientation, the technical efficiency of a firm is the maximum proportional (radial) reduction in all inputs needed to produce its current outputs in a specific technology.
- In output orientation, the technical efficiency of a firm is the maximum proportional (radial) expansion in all outputs possible from its current inputs in a specific technology.

Based on the above definition, between two firms $A(x_A, y_A)$ and $B(x_B, y_B)$, one can say that firm A is more efficient than firm B if it uses no more inputs to generate no fewer products and in at least one input or output it is strictly better.

In what follows, we apply the definition of "dominance" that is directly related to the notion of technical efficiency.

Definition 1.3. Dominance: Suppose x and x' are two m-vectors of inputs and y and y' are two s-vectors of outputs. $\begin{pmatrix} -x \\ y \end{pmatrix}$ dominates $\begin{pmatrix} -x' \\ y' \end{pmatrix}$ if and only if $\begin{pmatrix} -x \\ y \end{pmatrix} \geq \begin{pmatrix} -x' \\ y' \end{pmatrix}$ and $\begin{pmatrix} -x \\ y \end{pmatrix} \neq \begin{pmatrix} -x' \\ y' \end{pmatrix}$.

Based on the definition of "dominance," it is easy to prove that efficient firms are those that cannot be dominated by other firms. So, for a given technology set T, efficiency can be defined as follows:

Definition 1.4. Efficiency: Firm $(x, y) \in T$ is efficient if and only if there is no $(x', y') \in T$ that dominates (x, y).

The Debreu–Farrell input-oriented measure of technical efficiency is given by

$$TE_I(x, y) = Min\{\theta : (\theta x, y) \in T\}.$$

Moreover, the Debreu–Farrell output-oriented measure of technical efficiency is given by

$$(TE_O(x, y))^{-1} = Max\{\varphi : (x, \varphi y) \in T\}.$$

These two technical efficiency measures are illustrated in Fig. 1.4. Here $P(x, y)$ is located below the frontier $y = f(x)$ and its input-efficiency is $\theta < 1$ and its output-efficiency is $\frac{1}{\varphi} < 1$.

It is easy to see that the Debreu–Farrell measures of efficiency (TE_I and TE_O) are invariant to changes in the units of measurement. Moreover, TE_O is weakly monotone decreasing in outputs and TE_I is weakly monotone decreasing in inputs.

We can also combine the two orientations by simultaneously contracting inputs and expanding outputs along a given direction $d = \begin{pmatrix} d_x \\ d_y \end{pmatrix}$ to reach the efficient frontier. The directional measure of technical efficiency is defined by

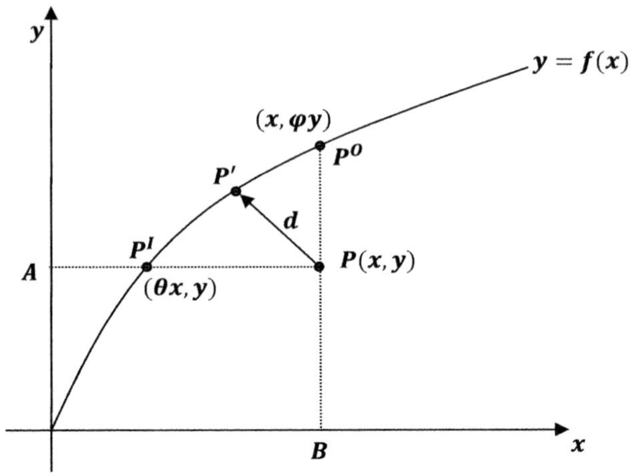

Fig. 1.4 Technical efficiency in three orientations

$$TE_D = Max\{\alpha : (x - \alpha d_x, y + \alpha d_y) \in T\}.$$

In the case of single-input and single-output, the moving procedure along the direction d is depicted in Fig. 1.4. In this case, point P is projected onto point P' on the efficient frontier.

1.4.2 (Input) Allocative Efficiency

In calculating technical efficiency, we have minimum information on inputs and outputs for each firm. Now, suppose that we have information on the firms' unit input prices $w = (w_1, \ldots, w_m) > 0$ and we are interested in minimizing the cost. The cost efficiency, $CE(x, y)$, is defined as the ratio of the minimum cost $C(y; w)$ to the actual cost, *i.e.*

$$CE(x, y) = \frac{C(y; w)}{\sum_{i=1}^{m} w_i x_i},$$

where $C(y; w)$ can be computed from the following cost minimization problem:

$$C(y; w) = \min_x \{w \cdot x : (x, y) \in T\}.$$

A measure of (input) allocative efficiency (IAE) is then derived as

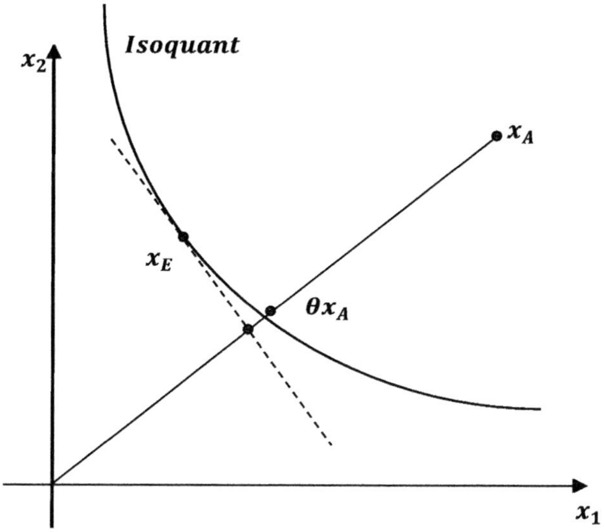

Fig. 1.5 Cost efficiency

$$IAE(x, y) = \frac{CE(x, y)}{TE_I(x, y)}.$$

We can then express the cost-efficiency measure of a firm as the product of its technical efficiency and (input) allocative efficiency measures, *i.e.*

$$CE(x, y) = TE_I(x, y) \cdot IAE(x, y).$$

It is thus clear that a firm is cost-efficient only when it is both technically and allocatively efficient, otherwise, the firm is cost-inefficient. The measurement of cost efficiency is illustrated in Fig. 1.5. In this figure, x_E represents the cost-minimizing input bundle of producing the output bundle y_A. So, the cost efficiency of firm A is computed as

$$CE_A = \frac{w \cdot x_E}{w \cdot x_A}.$$

As Fig. 1.5 shows, the measure of input technical efficiency of firm A is θ and, hence, the (input) allocative efficiency of this firm is computed as

$$IAE_A = \frac{w.x_E}{w.(\theta x_A)}.$$

As a result, we have the following CE decomposition:

$$IAE_A \times TE_I = CE_A.$$

This means that a firm is cost-efficient when the inputs are used in a technically efficient way with a minimal cost combination.

1.4.3 (Output) Allocative Efficiency

Now, suppose that the unit output prices are given as $p = (p_1, \ldots, p_s) > 0$ and the firm seeks to maximize its total revenue. The revenue efficiency $RE(x, y)$ is defined as the ratio of the actual revenue to the potential revenue $R(x; p)$, *i.e.*

$$RE(x, y) = \frac{p \cdot y}{R(x; p)},$$

where $R(x; p)$ can be obtained from the optimal solution of the following revenue-maximization problem:

$$R(x; p) = \max_{y} \{p.y : (x, y) \in T\}.$$

The measure of (output) allocative efficiency (OAE) is then derived as

$$OAE(x, y) = \frac{RE(x, y)}{TE_O(x, y)}.$$

Figure 1.6 illustrates the measurement of revenue efficiency. Here, the output bundle y_E maximizes the total revenue from the input bundle x_B. The revenue efficiency of firm B is then computed as

Fig. 1.6 Revenue efficiency

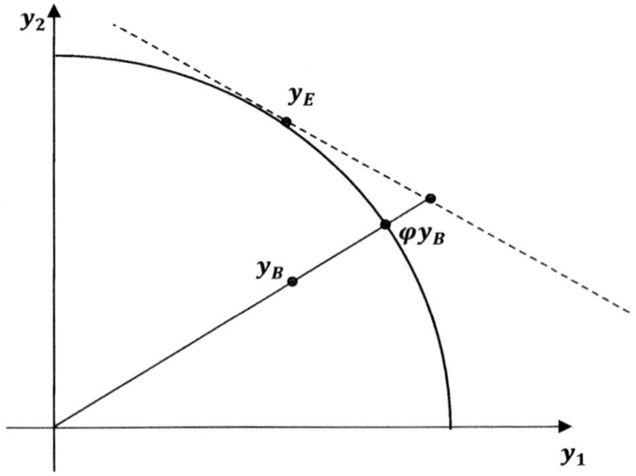

$$RE_B = \frac{p{\cdot}y_B}{p{\cdot}y_E}.$$

As one can see, since the output *TE* measure of firm *B* is $1/\varphi$, its OAE is computed as

$$OAE_B = \frac{p{\cdot}(\varphi y_B)}{p{\cdot}y_E}.$$

So, the resulting RE decomposition is expressed as

$$RE_B = \frac{p{\cdot}y_B}{p{\cdot}(\varphi y_B)} \cdot \frac{p{\cdot}(\varphi y_B)}{p{\cdot}y_E} = TE_o.OAE_B.$$

This shows that a firm is revenue efficient if it is both technically and allocatively efficient.

References

Banker, R. D., Charnes, A., & Cooper, W. W. (1984). Some models for estimating technical and scale inefficiencies in data envelopment analysis. *Management Science, 30*(9), 1078–1092.

Charnes, A., Cooper, W. W., & Rhodes, E. (1978). Measuring the efficiency of decision making units. *European Journal of Operational Research, 2*(6), 429–444.

Farrell, M. J. (1957). The measurement of productive efficiency. *Journal of the Royal Statistical Society: Series A (General), 120*(3), 253–281.

Koopmans, T. C. (1951). *Activity analysis of production and allocation*. Cowles Commission Monograph No. 13. New York/London: Wiley/Chapman and Hall.

Chapter 2
An Introduction to Data Envelopment Analysis

Following the seminal work of Farrell (1957), Charnes et al. (1978) introduced DEA as a deterministic and nonparametric efficiency evaluation tool. DEA is a linear programming-based technique that has been widely accepted as a competing methodology to evaluate the relative efficiency of entities or decision-making units, DMUs (Charles et al., 2016, 2018; Tsolas et al., 2020). DEA is a data-oriented technique (Zhu, 2020) that is used to construct an empirical production frontier to measure efficiency. Note that the original DEA program of Charnes et al. (1978) is based on the CRS specification of technology and is used to measure the *technical and scale efficiency* of DMUs. However, Banker et al. (1984) extended this program to the case of VRS to estimate purely *technical efficiency*. Over the past three decades, DEA has been widely used to evaluate the relative efficiency of production firms, the nature of the returns-to-scale, and the productivity changes. The DEA literature has seen a wide variety of applications across a plethora of domains, having become a powerful management science tool (Charles et al., 2018). In this chapter, we briefly review the fundamental concepts in DEA, along with the basic technologies and programs.

2.1 Symbols and Notations

In the DEA literature, the terms production units, firms, and DMUs are all used interchangeably. A short definition of a production unit or firm or DMU is that of an entity responsible for consuming inputs to generate outputs. Suppose there are j DMUs to be evaluated. An important problem before choosing a DEA model is the choice of inputs and outputs. If DEA is used as a benchmarking tool, then the inputs are "the best—the better" types of measures and the outputs are "the more—the better" types of measures. However, if undesirable factors are present in the production process, then this does not hold, and researchers must be careful in

A. Amirteimoori et al., *Stochastic Benchmarking*, International Series in Operations Research & Management Science 317,
https://doi.org/10.1007/978-3-030-89869-4_2

classifying the input–output measures. The following symbols and notations are used in this and subsequent chapters:

$$DMU_j : j = 1, \ldots, n : \text{the } j^{th} \; DMU;$$

$$DMU_o : \text{the } DMU \text{ under consideration for evaluation;}$$

$$x_j = \left(x_{1j}, \ldots, x_{mj}\right) : \text{the input vector of } DMU_j;$$

$$y_j = \left(y_{1j}, \ldots, y_{sj}\right) : \text{the output vector of } DMU_j;$$

$$U = (u_1, \ldots, u_s) : \text{the vector of output weights;}$$

$$V = (v_1, \ldots, v_m) : \text{the vector of input weights;}$$

$$P = (p_1, \ldots, p_s) : \text{the vector of unit output prices;}$$

$$C = (c_1, \ldots, c_m) : \text{the vector of unit input prices.}$$

2.2 Technology (Production Possibility) Set

As stated in the previous section, benchmarking uses a real production technology set and a utility function to evaluate the efficiency of the DMUs. The real technology (production possibility) set is usually unknown but can be empirically estimated with the help of some production axioms. Suppose there are j DMUs and each DMU_j $(j = 1, \ldots, n)$ uses the input vector $x_j = (x_{1j}, \ldots, x_{mj})$ to produce the output vector $y_j = (y_{1j}, \ldots, y_{sj})$. Each DMU_j is represented here by its input–output vector (x_j, y_j).

Generally, the technology set T is defined as

$$T = \left\{(x, y) : x \in \mathbb{R}_m^+, y \in \mathbb{R}_s^+, \text{the output } y \text{ can be produced by input } x\right\}.$$

In what follows, we introduce some axioms that are used to construct different technology sets. For more details, see Banker (1984), Banker et al. (1984), and Banker and Thrall (1992).

Axiom 1 *Inclusion of observations*: $(x_j, y_j) \in T$ for all $j = 1, \ldots, n$.

Axiom 2 *Convexity*: Any convex combination of feasible production possibilities is feasible. Mathematically, if $(x, y) \in T$ and $(x', y') \in T$, then $\lambda(x, y) + (1 - \lambda)(x', y') \in T$ for $\lambda \in [0, 1]$.

Axiom 3 *Different returns-to-scale possibilities*: A production possibility can be scaled up or down by any given positive scalar:

(i) if $(x_j, y_j) \in T$, then $\alpha(x_j, y_j) \in T$ for $\alpha > 0$ (CRS);
(ii) if $(x_j, y_j) \in T$, then $\alpha(x_j, y_j) \in T$ for $0 < \alpha \leq 1$ (non-increasing returns-to-scale, NIRS);
(iii) if $(x_j, y_j) \in T$, then $\alpha(x_j, y_j) \in T$ for $\alpha > 1$ (non-decreasing returns-to-scale, NDRS).

Axiom 4 *Free disposability*: Inputs and outputs are freely (strongly) disposable (assuming that inputs and outputs are all desirable). Mathematically, if $(x, y) \in T$, then for $x \leq x'$ and $y \geq y'$, $(x', y') \in T$.

Axiom 5 *Minimal extrapolation*: T is the intersection of all T' that satisfy Axioms: $1-4$.

Based on the above-stated axioms, five different empirical technology sets are constructed. In general form, a unique production possibility set, using Axioms: $1-5$ can be represented as

$$T^{(p)} = \left\{ (x, y) : \exists \lambda = (\lambda_1, \ldots, \lambda_n) \in S(p) : x \geq \sum_{j=1}^{n} \lambda_j x_j, \quad \sum_{j=1}^{n} \lambda_j y_j \geq y \right\},$$

and $p \in \{CRS, \ VRS, \ NIRS, \ NDRS, \ FDH\}$ and

$$S(CRS) = \{(\lambda_1, \ldots, \lambda_n) : \lambda_j \geq 0, \ j = 1, \ldots, n\};$$

$$S(VRS) = \left\{ (\lambda_1, \ldots, \lambda_n) : \sum_{j=1}^{n} \lambda_j = 1 \& \lambda_j \geq 0, \ j = 1, \ldots, n \right\};$$

$$S(NIRS) = \left\{ (\lambda_1, \ldots, \lambda_n) : \sum_{j=1}^{n} \lambda_j \leq 1 \& \lambda_j \geq 0, \ j = 1, \ldots, n \right\};$$

$$S(NDRS) = \left\{ (\lambda_1, \ldots, \lambda_n) : \sum_{j=1}^{n} \lambda_j \geq 1 \& \lambda_j \geq 0, \ j = 1, \ldots, n \right\};$$

$$S(FDH) = \left\{ (\lambda_1, \ldots, \lambda_n) : \sum_{j=1}^{n} \lambda_j = 1 \& \lambda_j \in \{0, 1\}, \ j = 1, \ldots, n \right\}.$$

where FDH represents free disposal hull.

Table 2.1 recapitulates the axioms that are used in each of these five technology sets.

Based on observed input–output data of six DMUs ($\boldsymbol{P_1}$, $\boldsymbol{P_2}$, $\boldsymbol{P_3}$, $\boldsymbol{P_4}$, $\boldsymbol{P_5}$, and $\boldsymbol{P_6}$), the different technology sets depicted in Fig. 2.1 are as follows:

- The half-line OC represents the CRS frontier;
- The piecewise linear frontier $AP_1P_2P_3P_4B$ represents the VRS frontier;
- The piecewise linear frontier $OP_2P_3P_4B$ represents the $NIRS$ frontier;

Table 2.1 Different technology sets

Technology	Axiom 1	Axiom 2	Axiom 3 (i)	Axiom 3 (ii)	Axiom 3 (iii)	Axiom 4	Axiom 5
T^{CRS}	✓	✓	✓	✓	✓	✓	✓
T^{VRS}	✓	✓	–	–	–	✓	✓
T^{IRS}	✓	✓	–	–	✓	✓	✓
T^{DRS}	✓	✓	–	✓	–	✓	✓
T^{FDH}	✓	–	–	–	–	✓	✓

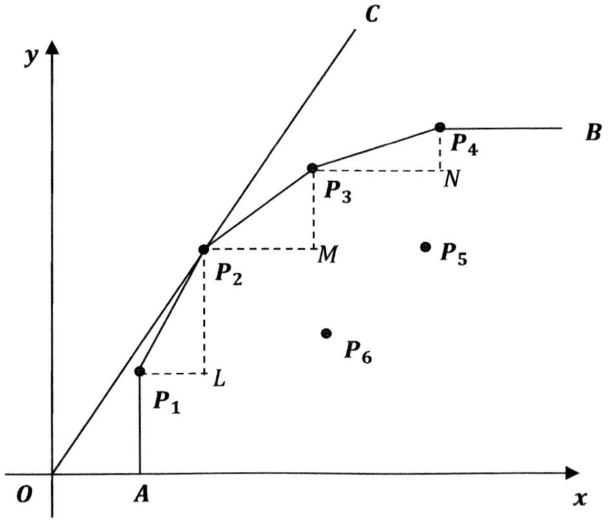

Fig. 2.1 Different technology sets under different returns-to-scale assumptions

- The piecewise linear frontier AP_1P_2C represents the *NDRS* frontier;
- The piecewise linear frontier $AP_1LP_2MP_3NP_4B$ represents the *FDH* frontier.

2.3 Basic DEA Programs

2.3.1 CCR Program

In this section, we deal with the first basic DEA program called the CCR program that was initiated by Charnes et al. (1978). The Debreu–Farrell input-oriented measure of the efficiency of DMU_o is set up as

$$TE_I^{(o)} = \text{Min}\left\{\theta : (\theta x_o, y_o) \in T^{(CRS)}\right\}, \tag{2.1}$$

where

$$T^{(CRS)} = \left\{(x, y) : x \geq \sum_{j=1}^{n} \lambda_j x_j, y \leq \sum_{j=1}^{n} \lambda_j y_j, \lambda_j \geq 0 : j = 1, \ldots, n\right\}.$$

Based on the algebraic formulation of $T^{(CRS)}$, we obtain the following linear program $CCR(o)$:

$$\begin{aligned} CCR(o) : TE_I^{(o)} &= \text{Min } \theta \\ \text{s.t.} \quad & \sum_{j=1}^{n} \lambda_j x_j \leq \theta x_o \\ & \sum_{j=1}^{n} \lambda_j y_j \geq y_o, \\ & \lambda_j \geq 0, \quad j = 1, \ldots, n. \end{aligned} \tag{2.2}$$

This is the envelopment form of the CCR model. The solution vector ($\theta = 1$, $\lambda_o = 1, \lambda_j = 0,\ j = 1, \ldots, n,\ j \neq o$) is feasible in model (2.2), and this guarantees that the optimal θ is not greater than unity.

To detect the input excesses and output shortfalls, the following two-step procedure is proposed:

Step 1: The linear program $CCR(o)$ is solved and suppose that θ^* is the optimal objective value.

Step 2: Solve the following linear program:

$$\begin{aligned} \text{Max } \Pi &= \sum_{i=1}^{m} d_i + \sum_{r=1}^{s} f_r \\ \text{s.t.} \quad & \sum_{j=1}^{n} \lambda_j x_{ij} + d_i = \theta^* x_{io}, \quad i = 1, \ldots, m, \\ & \sum_{j=1}^{n} \lambda_j y_{rj} - f_r = y_{ro}, \quad r = 1, \ldots, s, \\ & \lambda_j, d_i, f_r \geq 0, \quad \forall\ i, j, r. \end{aligned} \tag{2.3}$$

This two-step procedure can be modeled into the following linear program:

$$CCR(o) : TE_I^{(o)} = \text{Min } \theta - \varepsilon \left[\sum_{i=1}^{m} d_i + \sum_{r=1}^{s} f_r\right] \tag{2.4}$$

$$\text{s.t.} \quad \sum_{j=1}^{n} \lambda_j x_{ij} + d_i = \theta^* x_{io}, \qquad i = 1, \ldots, m,$$

$$\sum_{j=1}^{n} \lambda_j y_{rj} - f_r = y_{ro}, \qquad r = 1, \ldots, s,$$

$$\lambda_j, d_i, f_r \geq 0, \quad \forall\ i, j, r.$$

where $\varepsilon > 0$ is a non-Archimedean quantity.

Definition 2.1 DMU_o is fully efficient if and only if in all optimal solutions, $\theta^* = 1$ and $d_i^* = 0 : i = 1, \ldots, m$ and $f_r^* = 0 : r = 1, \ldots, s$. Otherwise, DMU_o is inefficient.

Suppose that DMU_o is inefficient. The reference unit for DMU_o is defined as

$$\overline{x}_o = \sum_{j=1}^{n} \lambda_j^* x_j,$$

$$\overline{y}_o = \sum_{j=1}^{n} \lambda_j^* y_j,$$

where λ_j^*s ($j = 1, \ldots, n$) are the optimal values obtained from program (2.4). Moreover, the reference set of DMU_o is given by

$$R_o = \left\{ j \in \{1, 2, \ldots, n\} : \lambda_j^* > 0 \right\}.$$

The very objective of finding a reference unit for the inefficient DMU_o is to see the extent of input excesses used, *i.e.*, $(x_o - \overline{x}_o)$ and output shortfalls made, *i.e.*, $(\overline{y}_o - y_o)$, based on which planning can be made to improve efficiency.

The dual formulation of program (2.2) is set up as

$$CCR(o) : TE_I^{(o)} = \text{Max} \quad U^t y_o \tag{2.5}$$

$$\text{s.t.} \quad V^t x_o = 1,$$

$$U^t y_j - V^t x_j \leq 0, \quad j = 1, \ldots, n,$$

$$(U^t, V^t) \geq 0,$$

where $U = (u_1, \ldots, u_s)^t$ and $V = (v_1, \ldots, v_m)^t$ are, respectively, the weights associated with the outputs and inputs of DMU_o. This is the multiplier form of the *CCR* program. The optimal input and output weights obtained from program (2.5) are most favorable to the DMU under evaluation so that it can obtain the maximum possible efficiency score.

Program (2.5) is the linear form of the following linear fractional programming problem proposed by Charnes et al. (1978):

$$CCR(o): TE_I^{(o)} = \text{Max} \ \frac{U^t y_o}{V^t x_o} \tag{2.6}$$

$$\text{s.t.} \quad \frac{U^t y_j}{V^t x_j} \le 1 \qquad j = 1, \ldots, n,$$

$$(U^t, V^t) \ge 0.$$

Program (2.6) can be easily transformed into the linear programming problem, *i. e.*, program (2.5), by using the Charnes–Cooper transformation method (Charnes & Cooper, 1963).

Similarly, the Debreu–Farrell output-oriented measure of the efficiency of DMU_o can be set up as

$$TE_O^{(o)} = \text{Max} \left\{ \phi : (x_o, \phi y_o) \in T^{(CRS)} \right\},$$

or in an equivalent form:

$$TE_O^{(o)} = \text{Max} \ \phi \tag{2.7}$$

$$\text{s.t.} \quad \sum_{j=1}^{n} \lambda_j x_{ij} \le x_o, \qquad i = 1, \ldots, m,$$

$$\sum_{j=1}^{n} \lambda_j y_{rj} \ge \phi y_o, \qquad r = 1, \ldots, s,$$

$$\lambda_j \ge 0 \quad j = 1, \ldots, n.$$

2.3.2 BCC Program

In the Debreu–Farrell efficiency program (2.1), if we replace $T^{(CRS)}$ with $T^{(VRS)}$, we obtain the following BCC program of Banker et al. (1984):

$$BCC(o): TE_I^{(o)} = \text{Min} \ \theta \tag{2.8}$$

$$\text{s.t.} \sum_{j=1}^{n} \lambda_j x_{ij} \leq \theta x_o, \qquad i = 1, \ldots, m,$$

$$\sum_{j=1}^{n} \lambda_j y_{rj} \geq y_o, \qquad r = 1, \ldots, s,$$

$$\sum_{j=1}^{n} \lambda_j = 1$$

$$\lambda_j \geq 0 \quad j = 1, \ldots, n$$

The main difference between the CCR and the BCC programs is the convexity constraint $\sum_{j=1}^{n} \lambda_j = 1$. The dual of program (2.8) is set up as

$$BCC(o) : TE_I^{(o)} = \text{Max} \quad U^t y_o - u_o \tag{2.9}$$

$$\text{s.t.} \quad U^t y_j - V^t x_j - u_o \leq 0, \qquad j = 1, \ldots, n,$$

$$V^t x_o = 1,$$

$$(U^t, V^t) \geq 0.$$

Program (2.9) is the linear form of the following linear fractional programming problem:

$$BCC(o) : TE_I^{(o)} = \text{Max} \quad \frac{U^t y_o - u_o}{V^t x_o} \tag{2.10}$$

$$\text{s.t.} \quad \frac{U^t y_j - u_o}{V^t x_j} \leq 1, \qquad j = 1, \ldots, n,$$

$$(U^t, V^t) \geq 0.$$

Definition 2.2. *DMU_o is said to be BCC-efficient if and only if there exists an optimal solution (U^*, V^*, u_o^*) to program (2.9) with $(U^*, V^*) > 0$ and $U^{*t} y_o - u_o^* = 1$.*

2.3.3 *Additive Program*

In previous subsections, the CCR and BCC programs were given in either input-orientation form (where the objective is to reduce inputs to become efficient) or output-orientation form (where the objective is to improve outputs to become efficient). However, efficiency can be achieved by maintaining both orientations simultaneously, *i.e.*, by reducing inputs and increasing outputs simultaneously. To this end, we can combine these two orientations in one single program, called the *additive program*. The mathematical formulation of the additive program under the CRS specification of technology is set up as

$$\text{Max} \sum_{i=1}^{m} s_i^- + \sum_{r=1}^{s} s_r^+ \tag{2.11}$$

$$\begin{aligned} \text{s.t.} \quad & \sum_{j=1}^{n} \lambda_j x_{ij} + s_i^- = x_{io}, \qquad i = 1, \ldots, m, \\ & \sum_{j=1}^{n} \lambda_j y_{rj} - s_i^+ = y_{ro}, \qquad r = 1, \ldots, s, \\ & \lambda_j, s_i^-, s_i^+ \geq 0, \quad \forall\ i, j, r. \end{aligned}$$

where s_i^- and s_r^+ are, respectively, the input surplus and output slack variables. This program is always feasible and bounded as well.

Definition 2.3. *DMU_o* is said to be additive-efficient if and only if $s_i^{-*} = 0$, $i = 1, \ldots, m$ and $s_r^{+*} = 0$, $r = 1, \ldots, R$, where s_i^{-*} , $i = 1, \ldots, m$ and s_r^{+*} , $r = 1, \ldots, s$ are the optimal solutions to program (2.11).

In this program, there is no priority on input reduction and output enhancement; it considers both input excess and output shortfall simultaneously to reach the efficiency frontier. It can easily be shown that DMU_o is additive-efficient in the program (2.11) if and only if it is BCC-efficient in program (2.9). In program (2.11), if DMU_o is found inefficient, the input and output slacks can be used to compute the target inputs and outputs as follows:

$$\begin{aligned} \overline{x}_{io} &= x_{io} - s_i^{-*}, \qquad i = 1, \ldots, m, \\ \overline{y}_{ro} &= y_{ro} + s_i^{+*}, \qquad r = 1, \ldots, s, \end{aligned} \tag{2.12}$$

The dual formulation of the additive program (2.11) can be set up as

$$\text{Max} \sum\nolimits_{i=1}^{m} v_i x_{io} + \sum\nolimits_{r=1}^{s} u_r y_{ro} \tag{2.13}$$

$$\begin{aligned} \text{s.t.} \quad & \sum\nolimits_{i=1}^{m} v_i x_{ij} + \sum\nolimits_{r=1}^{s} u_r y_{rj} \geq 0, \ j = 1, \ldots, n, \\ & v_i \geq 1, \qquad i = 1, \ldots, \text{m}, \\ & u_r \geq 1, \qquad r = 1, \ldots, \text{s}, \end{aligned}$$

In this dual formulation, if w_{io} and p_{ro} are, respectively, considered to be the unit price of the ith input and r^{th} output of DMU_o, then we have

$$\begin{aligned}\sum_{r=1}^{s} p_{ro} s_r^+ + \sum_{i=1}^{m} w_{io} s_i^- &= \sum_{r=1}^{s} p_{ro}(\bar{y}_{ro} - y_{ro}) + \sum_{i=1}^{m} w_{io}(x_{io} - \bar{x}_{io}) \\ &= \left(\sum_{r=1}^{s} p_{ro}\bar{y}_{ro} - \sum_{i=1}^{m} w\bar{x}_{io}\right) - \left(\sum_{r=1}^{s} p_{ro} y_{ro} - \sum_{i=1}^{m} w_{io} x_{io}\right).\end{aligned}$$

The last expression in the above equation represents the difference between the two profits: the first term is the profit made on the projection activity $(\bar{x}_o, \ \bar{y}_o)$ and the second term is the profit made on the observed activity $(\bar{x}_o, \ \bar{y}_o)$. This difference in profit shows the loss in profit due to technical inefficiency. This allows us to define an alternative measure of allocative efficiency by using the following program:

$$\begin{aligned} &\text{Max} \ \sum_{r=1}^{s} p_{ro} s_r^+ + \sum_{i=1}^{m} w_{io} s_i^- \\ \text{s.t.} \quad &\sum_{j=1}^{J} \lambda_j x_{ij} + s_i^- = x_{io}, \quad i = 1, \ldots, m, \\ &\sum_{j=1}^{J} \lambda_j y_{rj} - s_r^+ = y_{ro}, \quad r = 1, \ldots, s, \\ &\lambda_j, s_i^-, s_r^+ \geq 0, \quad \forall i, j \text{ and } r. \end{aligned} \tag{2.14}$$

2.3.4 Allocative Efficiency Programs

In Chap. 1, input and output allocative efficiencies were introduced in general form, based on their underlying technology set. When we use the DEA technology set, the resulting mathematical programming problems become linear programming problems. For example, if we use the CRS technology set $T^{(CRS)}$ to compute the (input) allocative efficiency (IAE) of DMU_o, we first solve the following cost-efficiency (CE) program:

$$\begin{aligned} &CE(x_o, y_o) = \text{Min} \ \frac{\sum_{i=1}^{m} w_{io} x_i}{\sum_{i=1}^{m} w_{io} x_{io}} \\ \text{s.t.} \quad &\sum_{j=1}^{n} \lambda_j x_{ij} \leq x_i, \quad i = 1, \ldots, m, \\ &\sum_{j=1}^{n} \lambda_j y_{rj} \geq y_{ro}, \quad r = 1, \ldots, s, \\ &\lambda_j, x_i \geq 0, \quad \forall j \text{ and } i, \end{aligned} \tag{2.15}$$

where $w_{io} : i = 1, \ldots, m$ are the unit input prices of DMU_o.

The (input) allocative efficiency of DMU_o can then be computed as the ratio of the CE to the (input) technical efficiency, *i.e.* $IAE(x_o, y_o) = CE(x_o, y_o)/TE_I^{(o)}$.

If we know the unit output prices, $p_{ro}: \ r = 1, \ldots, s$, the (output) allocative efficiency (OAE) of DMU_o can be computed. To do so, we first solve the following revenue-maximization problem:

$$\text{Max } \sum_{r=1}^{s} p_{ro} y_r \tag{2.16}$$

$$\text{s.t.} \quad \sum_{j=1}^{n} \lambda_j x_{ij} \leq x_{io}, \quad i = 1, \ldots, m,$$

$$\sum_{j=1}^{n} \lambda_j y_{rj} \geq y_r, \quad r = 1, \ldots, s,$$

$$\lambda_j, y_r \geq 0, \ \forall j \text{ and } r.$$

Assuming $y_r^* s$ $(r = 1, \ldots, s)$ to be the optimal solution values of program (2.16), the revenue efficiency of DMU_o is calculated as the ratio of the actual revenue to the potential revenue, *i.e.* $RE(x_o, y_o) = \sum_{r=1}^{s} p_{ro} y_{ro} / \sum_{r=1}^{s} p_{ro} y_r^*$. The output allocative efficiency can then be derived as the ratio of revenue efficiency to the (output) technical efficiency, *i.e.* $OAE(x_o, y_o) = RE(x_o, y_o)/TE_O^{(o)}$.

2.4 Returns-to-Scale

The concepts of returns-to-scale and economies of scale are often used synonymously in the literature. Either of these two concepts is generally taken to be a satisfactory way of treating scale economies without bothering much about whether they represent the same causal factors. However, although the concept of economies of scale is related to production size, this concept is much broader as it encompasses not only the RTS, but also variations in the input–output ratio due to factors such as indivisibility of production factors, the existence of external and internal economies, nature of contracts, and interrelationships between the organization and its stakeholders (Sahoo et al., 1999; Tone & Sahoo, 2003; Sahoo & Tone, 2013; Sahoo & Sengupta, 2014; Pérez-López et al., 2018).

In the economic theory of production, the quantitative change in the outputs of an efficient firm resulting from a unit proportionate change in all the inputs is referred to as returns-to-scale (RTS). In the case of a single-input-single-output technology characterized by the production function $y = f(x)$, the RTS or scale elasticity (SE) of a firm can be defined as

$$SE(x, y) = \frac{\text{Percentage change in output}}{\text{Unit percentage change in input}}$$

$$= \frac{\Delta y/y}{\Delta x/x} = \frac{\Delta y/\Delta x}{y/x} = \frac{MP}{AP},$$

where AP and MP represent, respectively, the average productivity and marginal productivity of the firm. Returns-to-scale are increasing, constant, and decreasing if $(\cdot) > 1$ ($MP > AP$), $SE(\cdot) = 1$ ($MP = AP$), and $SE(\cdot) < 1$ ($MP < AP$), respectively.

In the case of multiple inputs and single-output technology characterized by the production function$y = f(x_1, x_2, \ldots, x_m)$, if all the inputs are scaled by a single factor θ, the SE is defined as

$$SE(\theta) = \frac{dy/y}{d\theta/\theta}.$$

Regarding the computation of SE in DEA, let us bring in our (input) technical efficiency program (2.8) and its dual, program (2.9):

Program (2.8)

$$TE_I^{(o)} = \text{Min } \theta$$

$$\text{s.t. } \sum_{j=1}^{n} \lambda_j x_{ij} \leq \theta x_o \ (\forall i),$$

$$\sum_{j=1}^{n} \lambda_j y_{rj} \geq y_o \ (\forall r),$$

$$\sum_{j=1}^{n} \lambda_j = 1,$$

$$\lambda_j \geq 0 \ (\forall j).$$

Program (2.9)

$$TE_I^{(o)} = \text{Max } \sum_{r=1}^{s} u_r y_{ro} - u_o$$

$$\text{s.t. } \sum_{r=1}^{s} u_r y_{rj} - \sum_{i=1}^{m} v_i x_{ij} - u_o \leq 0 \ (\forall j),$$

$$\sum_{i=1}^{m} v_i x_{io} = 1,$$

$$u_r, v_i \geq 0 \ (\forall r, i).$$

Based on program (2.9) for firm o, we derive its transformation function as

$$\psi(\theta x_o, y_o) = \sum_{r=1}^{s} u_r\, y_{ro} - \sum_{i=1}^{m} v_i(\theta x_{io}) - u_0 = 0. \tag{2.17}$$

Assume that the transformation function (2.17) is differentiable. We can define the SE of firm o as

$$SE(\theta x_o, y_o) = -\frac{\sum_{r=1}^{s} \frac{\partial \psi(\theta x_o, y_o)}{\partial y_{ro}} y_{ro}}{\partial \psi\left(\frac{(\theta x_o, y_o)}{\partial(\theta x_{io})} x_o\right.}$$

$$= -\frac{\sum_{r=1}^{s} u_r y_{ro}}{-\sum_{i=1}^{m} v_i x_{io}} = 1 + u_0. \tag{2.18}$$

Firm o exhibits increasing RTS if $u_0 > 0$, CRS if $u_0 = 0$, and decreasing RTS if $u_0 < 0$.

2.5 Network DEA

As Shephard and Fare (1975) stated, some production systems may be considered as the joint, interacting action of a finite number of subsystems. Such a production process is referred to as network-structured production system. A simple network system is a two-stage production system as shown in Fig. 2.2 in which the first stage is fed by $x_{ij} : i = 1, \ldots, m$ to produce the intermediate measures $z_{dj} : d = 1, \ldots, D$. These intermediate measures are used as inputs to the second stage. The final outputs of the system are $y_{rj} : r = 1, \ldots, s$.

Two basic models have been developed to measure the relative efficiency of network systems: non-cooperative and cooperative models. In the following, these two approaches are introduced.

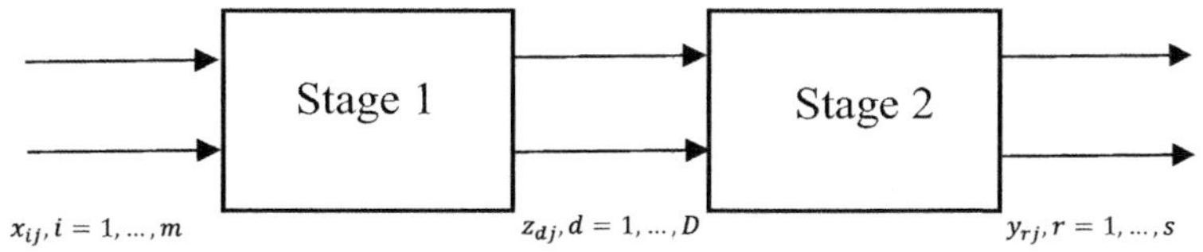

Fig. 2.2 Two-stage process for DMU_j

2.5.1 Non-cooperative Models (Leader–Follower)

In this section, the leader–follower approach is introduced to analyze the efficiency of the two-stage process depicted in Fig. 2.2. In non-cooperative game (Stackelberg or leader–follower game), there is a preference regarding the leader and the follower, where the importance of the leader is greater than that of the follower. So, if we assume the first stage is the leader, we first calculate the efficiency of the first stage. In CRS environment, the following program is used to optimize the efficiency of the first stage of DMU_o:

$$
\begin{aligned}
e_o^{\text{leader}*} &= \text{Max} \sum_{d=1}^{D} w_d z_{do} \\
\text{s.t.} \quad & \sum_{d=1}^{D} w_d z_{dj} - \sum_{i=1}^{m} v_i x_{ij} \leq 0 \quad j = 1, 2, \ldots, n, \\
& \sum_{i=1}^{m} v_i x_{io} = 1 \\
& w_d, v_i, \geq 0 \ (\forall d, i)
\end{aligned}
\tag{2.19}
$$

Suppose $e_o^{\text{follower}*}$ is the optimal value to program (2.19). When the efficiency of stage 1 is calculated, we come to optimize the relative efficiency of the second stage. To this end, we use the following program:

$$
\begin{aligned}
e_o^{\text{follower}*} &= \text{Max} \sum_{r=1}^{s} u_r y_{ro} \\
\text{s.t.} \quad & \sum_{r=1}^{s} u_r y_{rj} - \sum_{d=1}^{D} w_d z_{dj} \leq 0 \quad j = 1, 2, \ldots, n, \\
& \sum_{d=1}^{D} w_d z_{dj} - \sum_{i=1}^{m} v_i x_{ij} \leq 0 \quad j = 1, 2, \ldots, n, \\
& \sum_{i=1}^{m} v_i x_{io} = 1, \\
& \sum_{d=1}^{D} w_d z_{do} = e_o^{\text{leader}*}, \\
& w_d, v_i, u_r \geq 0 \ (\forall d, i, r),
\end{aligned}
\tag{2.20}
$$

2.5.2 Centralized Model

Now, we introduce the cooperative model. In this case, there is no preference regarding the two stages and both stages compete to optimize their efficiencies and at the same time, the performance of the whole system is maximized. Liang et al. (2006) have used the concept of cooperative game theory (or centralized control) in such a situation. Let us define the efficiencies of the first stage $\left(e_j^1\right)$ and second stage $\left(e_j^2\right)$, for each DMU_j, using the CRS-DEA model as follows:

$$e_j^1 = \frac{\sum_{d=1}^{D} w_d z_{dj}}{\sum_{i=1}^{m} v_i x_{ij}} \quad \text{and} \quad e_j^2 = \frac{\sum_{r=1}^{s} u_r y_{rj}}{\sum_{d=1}^{D} \widetilde{w}_d z_{dj}}, \tag{2.21}$$

in which $v_i, w_d, \widetilde{w}_d,$ and u_r are unknown non-negative weights. Now, what is important is the form of the efficiency of the whole system ($e_o^{\text{centralized}}$) as a function of e_j^1 and e_j^2. In some literature, it has been assumed that $w_d = \widetilde{w}_d : d = 1, \ldots, D$. In this case, $e_j^1.e_j^2 = \frac{\sum_{r=1}^{s} u_r y_{ro}}{\sum_{i=1}^{m} v_i x_{io}}$, that is, the weighted sum of final outputs to the weighted sum of initial inputs. In this section, we have also assumed that $e_o^{\text{centralized}} = e_j^1.e_j^2 = \frac{\sum_{r=1}^{s} u_r y_{ro}}{\sum_{i=1}^{m} v_i x_{io}}$. In this sense, the relative efficiency of the whole system is optimized by solving the following program:

$$\begin{aligned} e_o^{\text{centralized}} &= \text{Max} \ \frac{\sum_{r=1}^{s} u_r y_{ro}}{\sum_{i=1}^{m} v_i x_{io}} \\ \text{s.t.} \ & \frac{\sum_{d=1}^{D} w_d z_{dj}}{\sum_{i=1}^{m} v_i x_{ij}} \leq 1 \\ & \frac{\sum_{r=1}^{s} u_r y_{rj}}{\sum_{d=1}^{D} \widetilde{w}_d z_{dj}} \leq 1 \\ & \frac{\sum_{r=1}^{s} u_r y_{rj}}{\sum_{i=1}^{m} v_i x_{ij}} \leq 1 \\ & w_d = \widetilde{w}_d \\ & w_d, v_i, u_r \geq 0 \ (\forall d, i, r). \end{aligned} \tag{2.22}$$

Model (2.22) is a linear fractional program and it can easily be converted into a linear format by using the Charnes and Cooper (1963) transformation. The linear multiplier form of model (2.22) is as follows:

$$e_o^{\text{centralized}} = \text{Max} \sum_{r=1}^{s} u_r y_{rj} \tag{2.23}$$

$$\text{s.t.} \sum_{r=1}^{s} u_r y_{rj} - \sum_{d=1}^{D} w_d z_{dj} \leq 0 \quad j = 1, 2, \ldots, n$$

$$\sum_{d=1}^{D} w_d z_{dj} - \sum_{i=1}^{m} v_i x_{ij} \leq 0 \quad j = 1, 2, \ldots, n$$

$$\sum_{r=1}^{s} u_r y_{rj} - \sum_{i=1}^{m} v_i x_{ij} \leq 0 \quad j = 1, 2, \ldots, n$$

$$\sum_{i=1}^{m} v_i x_{io} = 1$$

$$w_d, v_i, u_r \geq 0 \ (\forall d, i, r).$$

Model (2.23) is a linear program and its dual formulation is as follows:

$$e_o^{\text{centralized}} = \text{Min } \theta \tag{2.24}$$

$$\text{s.t.} \sum_{j=1}^{n} (\lambda_j + \gamma_j) x_{ij} \leq \theta x_{io}, \qquad i = 1, \ldots, m,$$

$$\sum_{j=1}^{n} (\lambda_j + \xi_j) y_{rj} \geq y_{ro}, \qquad r = 1, \ldots, s,$$

$$\sum_{i=1}^{m} (\gamma_j - \xi_j) z_{dj} \geq 0, \qquad d = 1, \ldots, D,$$

$$\lambda_j, \gamma_j, \xi_j \geq 0, \qquad j = 1, \ldots, n.$$

References

Banker, R. D. (1984). Estimating most productive scale size using data envelopment analysis. *European Journal of Operational Research, 17*(1), 35–44.

Banker, R. D., & Thrall, R. M. (1992). Estimation of returns to scale using data envelopment analysis. *European Journal of Operational Research, 62*(1), 74–84.

Banker, R. D., Charnes, A., & Cooper, W. W. (1984). Some models for estimating technical and scale inefficiencies in data envelopment analysis. *Management Science, 30*(9), 1078–1092.

Charles, V., Peretto, C., & Gherman, T. (2016). Technical efficiency analytics for the public banks in Argentina. *ICPE Public Enterprise Half-Yearly Journal, 22*(1), 118–139.

Charles, V., Tsolas, I. E., & Gherman, T. (2018). Satisficing data envelopment analysis: A Bayesian approach for peer mining in the banking sector. *Annals of Operations Research, 269*(1–2), 81–102.

Charnes, A., Cooper, W. W., & Rhodes, E. (1978). Measuring the efficiency of decision making units. *European Journal of Operational Research, 2*(6), 429–444.

Charnes, A., & Cooper, W. W. (1963). Deterministic equivalents for optimizing and satisficing under chance constraints. *Operations Research, 11*(1), 18–39.

Farrell, M. J. (1957). The measurement of productive efficiency. *Journal of the Royal Statistical Society: Series A (General), 120*(3), 253–281.

Liang, L., Yang, F., Cook, W. D., & Zhu, J. (2006). DEA models for supply chain efficiency evaluation. *Annals of Operations Research, 145*(1), 35–49.

Pérez-López, G., Prior, D., & Zafra-Gómez, J. L. (2018). Temporal scale efficiency in DEA panel data estimations. An application to the solid waste disposal service in Spain. *Omega, 76*, 18–27.

Sahoo, B. K., & Tone, K. (2013). Non-parametric measurement of economies of scale and scope in non-competitive environment with price uncertainty. *Omega, 41*(1), 97–111.

Sahoo, B. K., Mohapatra, P. K., & Trivedi, M. L. (1999). A comparative application of data envelopment analysis and frontier translog production function for estimating returns to scale and efficiencies. *International Journal of Systems Science, 30*(4), 379–394.

Sahoo, B., & Sengupta, J. (2014). Neoclassical characterization of returns to scale in nonparametric production analysis. *Journal of Quantitative Economics, 12*(1), 77–85.

Shephard, R. W., & Fare, R. (1975). *A dynamic theory of production correspondences*. CALIFORNIA UNIV BERKELEY OPERATIONS RESEARCH CENTER.

Tone, K., & Sahoo, B. K. (2005). Scale, indivisibilities and production function in data envelopment analysis. *International Journal of Production Economics, 84*(2), 165–192.

Tsolas, I. E., Charles, V., & Gherman, T. (2020). Supporting better practice benchmarking: A DEA-ANN approach to bankbranch performance assessment. *Expert Systems with Applications, 160*, 113599.

Zhu, J. (2020). DEA under big data: Data enabled analytics and network data envelopment analysis. *Annals of Operations Research*, 1–23.

Chapter 3
Probability Theory

Probability theory is a branch of mathematical science that deals with the mathematical analysis of random events. Probability is commonly used to describe the mind's attitude to statements that we are not sure of. Statements usually take the form of "Will a particular event occur?" and the attitude of our minds will be of the form "How confident are we that this event will occur?" Our confidence can be described numerically, which is a value between 0 and 1 that we call *probability*. The more likely to occur an event is, the more confident we are that it will occur. The focus in this chapter is mainly on probability space, random variable, multi-dimensional probability distribution, expected value, variance, and covariance.

3.1 Probability Space

In this section, we present some important definitions that form the basis of probability theory.

Definition 3.1 A random experiment is an experiment whose outcome is not known in advance.

Definition 3.2 An event that may or may not occur as a result of a random experiment is called a random event.

For example, tossing a coin is a random experiment because its outcome is not known in advance. Coming heads is a random event because it may or may not happen.

Definition 3.3 The set of all possible outcomes of a random experiment associated with the phenomenon is called the *sample space*, represented by the symbol Ω. An individual element ω of Ω is called a *sample point*.

A. Amirteimoori et al., *Stochastic Benchmarking*, International Series in Operations Research & Management Science 317,
https://doi.org/10.1007/978-3-030-89869-4_3

Definition 3.4 The sample space Ω is discrete if Ω is a finite set or a countably infinite set.

Definition 3.5 The sample space Ω is continuous if Ω is an infinite set.

Example 3.1
(i) Coin tossing: $\Omega = \{H, T\}$.
(ii) Rolling one dice: $\Omega = \{1, 2, 3, 4, 5, 6\}$.
(iii) Picking one card at random in a pack of 52: $\Omega = \{1, 2, 3, \ldots, 52\}$.
(iv) An integer-valued random outcome: $\Omega = \{0, 1, 2, \ldots\}$.
(v) A non-negative, real-valued outcome: $\Omega = R_{+}$.
(vi) A random continuous parameter (such as time, weather, price or wealth, temperature, ...): $\Omega = R$.

An event is a collection of outcomes, which is represented by a subset of Ω. We can considera class $\mathcal{F}$ of events, *i.e.*, a class $\mathcal{F}$ of subsets of Ω [not necessarily all of the power sets of Ω, $\mathcal{P}(\Omega)$] as a $\sigma - algebra$ according to the following definition:

Definition 3.6 A collection $\mathcal{F}$ of events is a σ-algebra if it satisfies the following conditions:

(i) $\varnothing \in \mathcal{F}$;
(ii) For all countable sequences $(A_n)_{n \geq 1}$ such that $A_n \in \mathcal{F}$, $n \geq 1$, we have $\bigcup_{n \geq 1} A_n \in \mathcal{F}$;
(iii) $A \in \mathcal{F} \Rightarrow (\Omega \backslash A) \in \mathcal{F}$.

Since A is an event, we expect its complement A' to be an event. On the other hand, Ω will also be an event because $\Omega \subseteq \Omega$, so we expect $\varnothing$ to be an event. Ω is called a "certain event" and $\varnothing$ is an "impossible event." Consequently, since $\mathcal{F}$ is a subset of all subsets of Ω, we call it an "event space."

Example 3.2 Rolling one dice : $\Omega = \{1, 2, 3, 4, 5, 6\}$.

The event $A = \{1, 3, 5\}$ corresponds to "The result of the experiment is an odd number."

$\mathcal{F} := \{\Omega, \varnothing, \{1, 3, 5\}, \{2, 4, 6\}\}$ defines a $\sigma - algebra$ on Ω which corresponds to the knowledge of parity of an integer picked at random from 1 to 6.

$G := \{\Omega, \varnothing, \{2, 4, 6\}, \{2, 4\}, \{6\}, \{1, 2, 3, 4, 5\}, \{1, 3, 5, 6\}, \{1, 3, 5\}\}$ defines a $\sigma - algebra$ on Ω which is bigger than $\mathcal{F}$ and corresponds to the parity information contained in $\mathcal{F}$, completed by the knowledge of whether the outcome is equal to 6 or not.

Definition 3.7 Probability Measure. A probability measure is a mapping $\mathbb{P} : \mathcal{F} \to [0, 1]$ that for each event $A \in \mathcal{F}$ assigns a value in the interval [0, 1], with the following three axioms:

(a) For any event $A \in \mathcal{F}$, $\mathbb{P}(A) \geq 0$;
(b) $\mathbb{P}(\Omega) = 1$;
(c) $\mathbb{P}\left(\bigcup_{n\geq 1} A_n\right) = \sum_{n\geq 1} \mathbb{P}(A_n)$, whenever $A_i \cap A_k = \varnothing$, $i \neq k$.

These principles were put forward by the Russian mathematician and statistician Andrey Kolmogorov in 1933 as the principles of probability function and have since been known as the *Kolmogorov Principles*. The triple $(\Omega, \mathcal{F}, \mathbb{P})$ is called a probability space. This setting is generally referred to as the Kolmogorov framework. Based on the *Kolmogorov Principles*, many theorems have been established for the probability function. In the following, some important theorems for the probability function are given:

Theorem 3.1 Let $(\Omega, \mathcal{F}, \mathbb{P})$ be a probability space. Then, we have

(a) The probability of the event $\varnothing$is zero, *i.e.* $\mathbb{P}(\varnothing) = 0$.
(b) The probability of the union of two disjoint events is the sum of their probabilities, *i.e.*, $\mathbb{P}(A \cup B) = \mathbb{P}(A) + \mathbb{P}(B)$, where A and B are two disjoint events.
(c) The probability value for the complement of an event A is equal to $\mathbb{P}(A') = 1 - \mathbb{P}(A)$.
(d) If event A is a subset of event B, then $\mathbb{P}(A) \leq \mathbb{P}(B)$.
(e) If C is an event created by the union of two events A and B, then $\mathbb{P}(C) = \mathbb{P}(A \cup B) = \mathbb{P}(A) + \mathbb{P}(B) - \mathbb{P}(A \cap B)$.

Definition 3.8 Let A and B be two events in a sample space Ω, where $\mathbb{P}(A) \neq 0$; then the conditional probability of B given A is $\mathbb{P}(B|A) = \frac{\mathbb{P}(A \bigcap B)}{\mathbb{P}(A)}$.

It follows immediately that if A and B are the two events in the sample space Ω, where $\mathbb{P}(A) \neq 0$, then $\mathbb{P}(A \bigcap B) = \mathbb{P}(A).\ \mathbb{P}(B|A)$. Moreover, given that $\mathbb{P}(B) \neq 0$, since $\mathbb{P}(A|B) = \frac{\mathbb{P}(A \bigcap B)}{\mathbb{P}(B)}$, then $\mathbb{P}(A \bigcap B) = \mathbb{P}(B).\ \mathbb{P}(A|B)$.

Definition 3.9 Two events A and B are independent if and only if $\mathbb{P}(A \bigcap B) = \mathbb{P}(A).\ \mathbb{P}(B)$.

In other words, two events A and B are independent if the occurrence or nonoccurrence of one event does not affect the probability of the occurrence or nonoccurrence of the other event. In many cases, the result of an experiment depends on what happened at different intermediate stages. The following theorem on the law of total probability deals with this issue.

Theorem 3.3 Law of Total Probability. Let the events $A_1, A_2, \ldots, A_n$ be events that constitute a partition of sample space Ω, where $\mathbb{P}(A_i) \neq 0$, for $i = 1, 2, \ldots, n$. Then, for any event B in Ω, $\mathbb{P}(B) = \sum_{i=1}^{n} \mathbb{P}(A_i)\mathbb{P}(B|A_i)$.

In probability theory and statistics, the Bayes theorem describes the probability of an event, based on prior knowledge of conditions that might be related to the event. For example, if the probability that someone has cancer is related to his/her age, then using the Bayes theorem, the age can be used to more accurately assess the probability of cancer than it can be done without the knowledge of the age.

Theorem 3.4 Let the events $A_1, A_2, \ldots, A_n$ constitute a partition on the sample space Ω, where $\mathbb{P}(A_i) \neq 0$, for $i = 1, 2, \ldots, n$. Then, for any event B in Ω with $\mathbb{P}(B) \neq 0$,

$$\mathbb{P}(A_r|B) = \frac{\mathbb{P}(A_r).\mathbb{P}(B|A_r)}{\sum_{i=1}^{n} \mathbb{P}(A_i)\mathbb{P}(B|A_i)} \qquad r = 1, \ldots, n$$

3.2 Random Variable

Definition 3.10 Let $(\Omega, \mathcal{F}, \mathbb{P})$ be a probability space. A real-valued random variable on this probability space is a measurable mapping $X : (\Omega, \mathcal{F}) \to (\mathbb{R}, \mathcal{B}(\mathbb{R}))$, *i.e.* $X^{-1}(G) \in \mathcal{F}$ for all $G \in \mathcal{B}(\mathbb{R})$.

Note that X is a random variable from a probability space Ω into the state space $\mathbb{R}$, which maps each $\omega \in \Omega$ to $X(\omega) \in \mathbb{R}$.

Definition 3.11 A random variable X is continuous when Ω is continuous and is discrete when Ω is discrete.

Example 3.3 When we roll two dice, $\Omega := \{1, 2, 3, 4, 5, 6\} \times \{1, 2, 3, 4, 5, 6\} = \{(1, 1), (1, 2), \ldots, (6, 6)\}$. Consider $\mathrm{X} : \Omega \to \mathbb{R}$ and $(\mathrm{k}, \mathrm{l}) \to \mathrm{k} + \mathrm{l}$. Then, X is a random variable that gives the sum of the two numbers appearing on each dice.

Definition 3.12 If X is a discrete random variable on $(\Omega, \mathcal{F}, \mathbb{P})$, the function f (x) = P(X = x) for each x in the range of X is called the probability distribution of X.

Proposition 3.1 Let X be a discrete random variable on $(\Omega, \mathcal{F}, \mathbb{P})$. Then, f is a probability distribution of X if and only if it satisfies the following conditions:

(a) $\mathrm{f(x)} \geq 0$, for each value of x;
(b) $\sum_{\mathrm{x}} \mathrm{f(x)} = 1$.

Example 3.4 (Freund et al., 2004) Check whether the function given by $\mathrm{f(x)} = \frac{\mathrm{x}+2}{25}$, for $\mathrm{x} = 1, 2, 3, 4, 5$, can serve as the probability distribution of a discrete random variable.

By introducing the possible values in the function, $\mathrm{f}(1) = \frac{3}{25}$, $\mathrm{f}(2) = \frac{4}{25}$, $\mathrm{f}(3) = \frac{5}{25}$, $\mathrm{f}(4) = \frac{6}{25}$, $\mathrm{f}(5) = \frac{7}{25}$, since these values are all non-negative, the first condition of proposition 3.1 is satisfied, and since $f(1)+f(2)+f(3)+f(4)+f(5)=\frac{3}{25}+\frac{4}{25}+\frac{5}{25}+\frac{6}{25}+\frac{7}{25}=1$, the second condition is also satisfied. Thus, the given function can serve as the probability distribution of a random variable having the range $\{1,2,3,4,5\}$.

Definition 3.13 Distribution Function. If X is a discrete random variable on $(\Omega, \mathcal{F}, \mathbb{P})$, the function $F_X(x)$ is defined by $F_X(x)=P(X\leq x)=\sum_{t\leq x} f(t),\ -\infty<x<\infty$ and is called the distribution function or the cumulative distribution of X, in which f(t) is the value of the probability distribution of X at t.

Proposition 3.2 Let F_X be a cumulative distribution function of a discrete random variable X. Then,

(a) F_X is a non-decreasing function;
(b) $F_X(-\infty)=0$ and $F_X(+\infty)=1$.

Definition 3.14 Probability density function. A function $f_X:\mathbb{R}\to\mathbb{R}_+$ is called a probability density function of the continuous random variable X if and only if $P(a\leq x\leq b)=\int_a^b f_X(x)dx$, for any real constants a and b with $a\leq b$.

The probability density functions are also referred to as probability densities, density functions, densities, or pdf.

Theorem 3.5 If X is a continuous random variable on $(\Omega, \mathcal{F}, \mathbb{P})$ and a and b are real constants with $a\leq b$, then $P(a\leq X\leq b)=P(a\leq X<b)=P(a<X\leq b)=P(a<X<b)$.

Proposition 3.3 Let X be a continuous random variable on $(\Omega, \mathcal{F}, \mathbb{P})$. Then f_X is a probability density of X if and only if it satisfies the following conditions:

(a) $f_X(x)\geq 0,\quad \text{for}\ -\infty<x<\infty$;
(b) $\int_{-\infty}^{\infty} f_X(x)dx=1$.

Example 3.5 (Freund et al., 2004) If X has the probability density:

$$f_X(x)=\begin{cases} k.e^{-3x}, \text{for } x>0; \\ 0 \qquad\quad, \text{otherwise}; \end{cases}$$

find k and $P(0.5\leq X\leq 1)$.

Solution: To satisfy the second condition of Proposition 3.3, we must have $\int_{-\infty}^{\infty} f_X(x)dx=\int_0^\infty k.e^{-3x}dx=k.\frac{e^{-3x}}{-3}\Big|_0^\infty=\frac{k}{3}=1$. It follows that $k=3$. Moreover, $P(0.5\leq X\leq 1)=\int_{0.5}^1 3e^{-3x}dx=-e^{-3x}\Big|_{0.5}^1=-e^{-3}+e^{-1.5}=0.173$.

Definition 3.15 Let X be a continuous random variable on $(\Omega, \mathcal{F}, \mathbb{P})$. The distribution function or the cumulative distribution function (CDF) of X, $F_X:\mathbb{R}\to[0,1]$, is defined by $F_X(x) \equiv P(X \leq x) = \int_{-\infty}^{x} f_X(t)dt$, for $x \in \mathbb{R}$, where $f_X(t)$ is the value of the probability density of X at t.

That is, $F_X(x)$ is the probability that the random variable X takes a value, which is less than or equal to x.

Proposition 3.4 Let F_X be a cumulative distribution function. Then:

(a) $0 \leq F_X(x) \leq 1$,
(b) F_X is a non-decreasing function,
(c) $F_X(-\infty) = \lim_{x \to -\infty} F_X(x) = 0$ and $F_X(+\infty) = \lim_{x \to +\infty} F_X(x) = 1$.

Theorem 3.6 If $f_X(x)$ and $F_X(x)$ are the probability density and cumulative distribution functions of the random variable X, respectively, then for any real constants a and b with $a \leq b$, we have $P(a < X \leq b) = F(b) - F(a)$ and $f_X(x) = \frac{dF_X(x)}{dx}$.

Remark 3.1 $P(x = a) = 0$.

Example 3.6 Find the distribution function of the random variable X in Example 3.6 and use it to re-evaluate $P(0.5 \leq X \leq 1)$.

For $X > 0$, $F_X(x) = \int_{-\infty}^{x} f_X(t)dt = \int_0^x 3e^{-3t}dt = -e^{-3t}|_0^x = 1 - e^{-3x}$ and since $F(x) = 0$, for $x \leq 0$, we can write

$$F(X) = \begin{cases} 0 & , \quad \text{for } x \leq 0; \\ 1 - e^{-3x}, & \quad \text{for } x > 0. \end{cases}$$

Now, to determine $P(0.5 \leq X \leq 1)$, we use Theorem 3.6. So, $P(0.5 \leq X \leq 1) = F(1) - F(0.5) = (1 - e^{-3}) - (1 - e^{-1.5}) = 0.173$.

In probability theory, a multivariate random variable or random vector is a list of random variables, the values of which are unknown. For example, while a given person has a specific age, height, and weight, the representation of these features for an unspecified person in a group would be a random vector. Similarly, for the random variable, we now define multivariate random variables, or random vectors, as multivariate functions.

Definition 3.16 An n-dimensional random variable or vector X is a (measurable) function from the probability space Ω to $\mathbb{R}^n$, *i.e.* $X : \Omega \to \mathbb{R}^n$.

Definition 3.17 Joint Probability Distribution. Let $X_1, X_2, \ldots, X_n$ be discrete random variables. The joint probability distribution of $X_1, X_2, \ldots, X_n$ is defined as $f_{X_1,X_2,\ldots,X_n}(x_1, x_2, \ldots, x_n) = P(X_1 = x_1, X_2 = x_2, \ldots, X_n = x_n)$, where $x_i \in$ Range (X_i), $i = 1, \ldots, n$.

Theorem 3.7 A multi-dimensional function can serve as the joint probability distribution of the discrete random variables $X_1, X_2, \ldots, X_n$ if and only if its values, $f_{X_1,X_2,\ldots,X_n}(x_1, x_2, \ldots, x_n)$, satisfy the following conditions:

(a) $f_{X_1,X_2,\ldots,X_n}(x_1, x_2, \ldots, x_n) \geq 0$ for each $(x_1, x_2, \ldots, x_n)$ within its domain;
(b) $\sum_{x_1}\sum_{x_2}\cdots\sum_{x_n} f_{X_1,X_2,\ldots,X_n}(x_1, x_2, \ldots, x_n) = 1$, where the multiple summations extend over all possible $x_1, x_2, \ldots, x_n$ within its domain.

Definition 3.18 Joint Cumulative Distribution. Let $X_1, X_2, \ldots, X_n$ be discrete random variables. The joint distribution function or the joint cumulative distribution of $X_1, X_2, \ldots, X_n$ is defined as

$$F_{X_1,X_2,\ldots,X_n}(x_1,x_2,\ldots,x_n) = P(X_1 \leq x_1, X_2 \leq x_2, \ldots, X_n \leq x_n)$$
$$= \sum_{t_1 \leq x_1}\sum_{t_2 \leq x_2}\cdots\sum_{t_n \leq x_n} f_{X_1,X_2,\ldots,X_n}(t_1,t_2,\ldots,t_n) \text{ for } x_1,x_2,\ldots,x_n \in \mathbb{R},$$

for all $(x_1, x_2, \ldots, x_n)$ within the range of $X_1, X_2, \ldots, X_n$.

Definition 3.19 Joint Probability Density Function. A function $f_{X_1,\ldots X_n} : \mathbb{R}^n \to \mathbb{R}_+$ is called a joint probability density function of the continuous random variables $X_1, X_2, \ldots, X_n$ if and only if $P(\{X_1, X_2, \ldots, X_n\} \in A) = \int\int\cdots\int f_{X_1,\ldots X_n}(x_1, x_2, \ldots, x_n)dx_1dx_2\ldots dx_n$, for any region A in the $\mathbb{R}^n$.

Theorem 3.8 Let $X_1, X_2, \ldots, X_n$ be continuous random variables on $(\Omega, \mathcal{F}, \mathbb{P})$. Then, $f_{X_1,\ldots X_n}$ is a joint probability density function of $X_1, X_2, \ldots, X_n$ if and only if it satisfies the following conditions:

(a) $f_{X_1,X_2,\ldots,X_n}(x_1, x_2, \ldots, x_n) \geq 0, \quad \text{for} \quad x_1, x_2, \ldots, x_n \in \mathbb{R}$;
(b) $\int_{-\infty}^{\infty}\int_{-\infty}^{\infty}\cdots\int_{-\infty}^{\infty} f_{X_1,X_2,\ldots,X_n}(x_1, x_2, \ldots, x_n)dx_1\, dx_2\ldots dx_n = 1$.

Definition 3.20 Joint Distribution Function. If $X_1, X_2, \ldots, X_n$ are continuous random variables on $(\Omega, \mathcal{F}, \mathbb{P})$, then the function is given by

$$F_{X_1,X_2,\ldots,X_n}(x_1,x_2,\ldots,x_n) = P(X_1 \leq x_1, X_2 \leq x_2, \ldots, X_n \leq x_n)$$
$$= \int_{-\infty}^{x_1}\int_{-\infty}^{x_2}\cdots\int_{-\infty}^{x_n} f_{X_1,X_2,\ldots,X_n}(t_1,t_2,\ldots,t_n)dt_1dt_2\ldots dt_n, \text{for } x_1,x_2,\ldots,x_n \in \mathbb{R}$$

is called the joint distribution function of $X_1, X_2, \ldots, X_n$, in which $f_{X_1,X_2,\ldots,X_n}(t_1, t_2, \ldots, t_n)$ is the joint probability density of $X_1, X_2, \ldots, X_n$ at $(t_1, t_2, \ldots, t_n)$.

Theorem 3.9
(a) $F_{X_1,X_2,\ldots,X_n}(-\infty, -\infty, \ldots, -\infty) = 0$;
(b) $F_{X_1,X_2,\ldots,X_n}(\infty, \infty, \ldots, \infty) = 1$;

(c) If $a_1 < b_1$, $a_2 < b_2$, …, $a_n < b_n$, then $F_{X_1,X_2,\ldots,X_n}(a_1, a_2, \ldots, a_n) \leq F_{X_1,X_2,\ldots,X_n}(b_1, b_2, \ldots, b_n)$;

(d) $f_{X_1,X_2,\ldots,X_n}(x_1, x_2, \ldots, x_n) = \frac{\partial^n}{\partial x_1 \ldots \partial x_n} F_{X_1,X_2,\ldots,X_n}(x_1, x_2, \ldots, x_n)$.

Theorem 3.10 Let $X_i : i = 1, \ldots, n$ be random variables with probability distributions $f_{X_i}(x_i), i = 1, \ldots, n$, and let $f_{X_1,X_2,\ldots,X_n}(x_1, x_2, \ldots, x_n)$ be the joint probability distribution function of $X_1, X_2, \ldots, X_n$. Then, $X_1, X_2, \ldots, X_n$ are independent if and only if $f_{X_1,X_2,\ldots,X_n}(x_1, x_2, \ldots, x_n) = f_{X_1}(x_1).f_{X_2}(x_2).\ldots.f_{X_n}(x_n)$, for all $(x_1, x_2, \ldots, x_n)$ within their range.

3.3 Mathematical Expectation

The expectation, or expected value, of a random variable X is the mean or average value of X. In practice, expectations can be even more useful than probabilities.

For example, bank deposits or the prices of inputs of the DMUs can be considered as random variables. In such cases, we usually refer to their expected values rather than their actual values.

Definition 3.21 Expected Value. If X is a discrete random variable and $f_X(x)$ is its probability distribution function, the expected value of X is $\mathbb{E}(X) = \sum_x x.f_X(x)$.

Correspondingly, if X is a continuous random variable and $f_X(x)$ is its probability density, the expected value of X is $\mathbb{E}(X) = \int_{-\infty}^{\infty} x.f_X(x)dx$.

Theorem 3.11 If X is a discrete random variable and $f_X(x)$ is its probability distribution function, the expected value of g(X), as a function of X, is given by $\mathbb{E}[g(X)] = \sum_x g(x).f_X(x)$. Correspondingly, if X is a continuous random variable and $f_X(x)$ is its probability density function, then the expected value of g(X), as a function of X, is given by $\mathbb{E}[g(X)] = \int_{-\infty}^{\infty} g(x).f_X(x)dx$.

Theorem 3.12 If a and b are constants, then $\mathbb{E}(aX + b) = a\mathbb{E}(X) + b$.

Theorem 3.13 If $c_1, c_2, \ldots, c_n$ are constants, then $\mathbb{E}\left[\sum_{i=1}^{n} c_i g_i(X)\right] = \sum_{i=1}^{n} c_i \mathbb{E}[g_i(X)]$.

Theorem 3.14 If $X_1, X_2, \ldots, X_n$ are independent random variables, then $\mathbb{E}(X_1 X_2 \ldots X_n) = \mathbb{E}(X_1)\mathbb{E}(X_2)\ldots\mathbb{E}(X_n)$.

Definition 3.22 Variance. Let X be a random variable with a finite expected value, μ. Then, the variance of X, denoted by σ^2, σ_X^2, or var(X), is defined by $\sigma^2 = \mathbb{E}\left[(X - \mu)^2\right]$.

The positive square root of the variance, σ, is called the standard deviation of X.

Theorem 3.15

$$\sigma^2 = \mathbb{E}(X^2) - \mathbb{E}^2(X).$$

Theorem 3.16 If X has the variance σ^2 and a and b are constants, then var $(aX + b) = a^2\sigma^2$.

Herein, we represent Chebyshev's inequality that enables us to derive bounds on probabilities when only the mean or both the mean and the variance of the probability distribution are known and valid for all distributions of each random variable.

Theorem 3.17 Chebyshev's inequality. If X is a random variable with finite mean μ and variance σ^2, then for any value $k > 0$, we have $\mathbb{P}(|X - \mu| \geq k) \leq \frac{\sigma^2}{k^2}$.

Definition 3.23 Covariance. Let X and Y be two random variables with finite expected values μ_X and μ_Y, respectively. Then, the covariance of X and Y, denoted by σ_{XY} or cov(X, Y), is defined by $\text{cov}(X, Y) = \mathbb{E}[(X - \mu_X)(Y - \mu_Y)]$.

Theorem 3.18

$$\text{cov}(X, Y) = \mathbb{E}(XY) - \mathbb{E}(X)\mathbb{E}(Y).$$

It can easily be shown that if X and Y are two independent random variables, then $\text{cov}(X, Y) = 0$.

Theorem 3.19 If $X_1, X_2, \ldots, X_n$ are random variables and $Y = \sum_{i=1}^{n} a_i X_i$, where $a_1, a_2, \ldots, a_n$ are constants, then $\text{var}(Y) = \sum_{i=1}^{n} a_i^2.\text{var}(X_i) + 2\sum\sum_{i<j} a_i a_j.\text{cov}(X_i, X_j)$.

Proof. See Freund et al. (2004).

Corollary 3.1 If $X_1, X_2, \ldots, X_n$ are independent random variables and $Y = \sum_{i=1}^{n} a_i X_i$, then $\text{var}(Y) = \sum_{i=1}^{n} a_i^2.\text{var}(X_i)$.

Theorem 3.20 If $X_1, X_2, \ldots, X_n$ are random variables and $Y_1 = \sum_{i=1}^{n} a_i X_i$ and $Y_2 = \sum_{i=1}^{n} b_i X_i$, where $a_1, a_2, \ldots, a_n, b_1, b_2, \ldots, b_n$ are constants, then $\text{cov}(Y_1, Y_2) = \sum_{i=1}^{n} a_i b_i.\text{var}(X_i) + \sum\sum_{i<j} (a_i b_j + a_j b_i).\text{cov}(X_i X_j)$.

Corollary 3.2 If the random variables $X_1, X_2, \ldots, X_n$ are independent, $Y_1 = \sum_{i=1}^{n} a_i X_i$ and $Y_2 = \sum_{i=1}^{n} b_i X_i$, then $\text{cov}(Y_1, Y_2) = \sum_{i=1}^{n} a_i b_i.\text{var}(X_i)$.

Definition 3.24 Correlation Coefficient. The correlation coefficient of two random variables X and Y, denoted by $\rho(X, Y)$, is defined by $\rho(X, Y) = \frac{\mathrm{cov}(X, Y)}{\sqrt{\mathrm{var}(X)}\sqrt{\mathrm{var}(Y)}}$.

It can be shown that $-1 \leq \rho(X, Y) \leq 1$. The correlation coefficient is a measure of the degree of linearity between X and Y. A value of $\rho(X, Y)$ near +1 or −1 indicates a high degree of linearity between X and Y, whereas a value near 0 indicates that such linearity is absent. A positive value of $\rho(X, Y)$ indicates that Y tends to increase when X does, whereas a negative value indicates that Y tends to decrease when X increases. If $\rho(X, Y) = 0$, then X and Y are said to be uncorrelated.

3.4 Discrete Distributions

In this section, we introduce some of the most used probability distributions. Due to the importance of normal and chi-square distributions, these two distributions are studied in more detail.

Definition 3.25 Discrete uniform distribution. A random variable X has a **discrete uniform distribution** and it is referred to as a discrete uniform random variable if and only if its probability distribution is given by $f(x) = \frac{1}{k}$, for $k = x_1, x_2, \ldots x_k$.

A discrete uniform random variable holds all its values with equal probabilities.

Definition 3.26 Bernoulli distribution. A random variable X has a **Bernoulli distribution** and it is referred to as a Bernoulli random variable if and only if its probability distribution is given by $f(x; p) = p^x(1 - p)^{1 - x}$, for $x = 0, 1$.

Theorem 3.21 Let X be a Bernoulli random variable. The mean and variance of X are then $\mu = p$ and $\sigma^2 = p(1 - p)$, respectively.

Definition 3.27 Binomial distribution. A random variable X has a binomial distribution and it is referred to as a binomial random variable if and only if its probability distribution is given by $b(x; n, p) = \binom{n}{x} p^x(1 - p)^{n-x}$, for $x = 0, 1, \ldots, n$.

Theorem 3.22 Let X be a Binomial random variable. The mean and variance of X are then $\mu = np$ and $\sigma^2 = np(1 - p)$, respectively.

Definition 3.28 Poisson distribution. A random variable X has a Poisson distribution and it is referred to as a Poison random variable if and only if its probability distribution is given by $p(x; \lambda) = \frac{\lambda^x e^{-\lambda}}{x!}$, for $x = 0, 1, 2, \ldots$, where X is the mean of successes in each given time interval or region.

Theorem 3.23 Let X be a Poisson random variable. The mean and variance of X are then $\mu = \lambda$ and $\sigma^2 = \lambda$, respectively.

3.5 Continuous Distributions

Definition 3.29 Continuous uniform distribution. A random variable X has a uniform distribution and it is referred to as a continuous uniform random variable if and only if its probability density is given by

$$f(x) = \begin{cases} \frac{1}{b-a}, & a \leq x \leq b; \\ 0 \quad , & \text{elsewhere.} \end{cases}$$

In other words, the random variable X is uniformly distributed at intervals.

Theorem 3.24 Let X be a uniform random variable. The mean and variance of X are then $\mu = \frac{a+b}{2}$ and $\sigma^2 = \frac{1}{12}(b-a)^2$, respectively.

Definition 3.30 Gamma function. The gamma function of α, denoted by $\Gamma(\alpha)$, is defined as

$$\Gamma(\alpha) = \int_0^{\infty} x^{\alpha-1} e^{-x} dx.$$

Corollary 3.3

$$\Gamma(\alpha) = (\alpha - 1)!$$

Definition 3.31 Gamma distribution. A random variable X has a gamma distribution and it is referred to as a gamma random variable if and only if its probability density function is given by

$$g(x; \alpha, \beta) = \begin{cases} \frac{1}{\beta^{\alpha}\Gamma(\alpha)} x^{\alpha-1} e^{-x/\beta}, & \text{for } x > 0; \\ 0 & \text{, elsewhere;} \end{cases}$$

where $\alpha > 0$ and $\beta > 0$.

Theorem 3.25 Let X be a gamma random variable. The mean and variance of X are then $\mu = \alpha\beta$ and $\sigma^2 = \alpha\beta^2$, respectively.

Definition 3.32 Exponential distribution. A random variable X has an exponential distribution and it is referred to as an exponential random variable if and only if its probability density is represented by

$$g(x,\theta)=\begin{cases}\dfrac{1}{\theta}e^{-x/\theta}, & \text{for } x>0;\\ 0\quad , & \text{elsewhere;}\end{cases}$$

where $\theta > 0$.

Note: The exponential distribution is a special case of gamma distribution with $\alpha = 1$ and $\beta = \theta$.

Theorem 3.26 Let X be an exponential random variable. The mean and variance of X are then $\mu = \theta$ and $\sigma^2 = \theta^2$, respectively.

Definition 3.33 Weibull distribution. A random variable X has a Weibull distribution and it is referred to as a Weibull random variable if and only if its probability density is represented by

$$f(x)=\begin{cases}kx^{\beta-1}e^{-\alpha x^{\beta}}, & \text{for } x>0;\\ 0 & \text{, elsewhere;}\end{cases}$$

where $\alpha > 0$ and $\beta > 0$.

Note: The exponential distribution is a special case of Weibull distribution with $\beta = 1$.

Definition 3.34 Beta distribution. A random variable X has **beta distribution** and it is referred to as a beta random variable if and only if its probability density is represented by

$$f(x;\alpha,\beta)=\begin{cases}\dfrac{\Gamma(\alpha+\beta)}{\Gamma(\alpha)\Gamma(\beta)}x^{\alpha-1}(1-x)^{\beta-1}, & \text{for } 0<x<1;\\ 0 & \text{, elsewhere;}\end{cases}$$

where $\alpha > 0$ and $\beta > 0$.

Theorem 3.27 Let X be a Beta random variable. The mean and variance of X are then $\mu = \frac{\alpha}{\alpha+\beta}$ and $\sigma^2 = \frac{\alpha\beta}{(\alpha+\beta)^2(\alpha+\beta+1)}$, respectively.

3.6 The Normal Distribution

In probability theory, the normal distribution is one of the most important statistical distributions. This distribution is sometimes referred to as the Gaussian distribution or the Laplace–Gauss distribution. In this section, we briefly summarize the properties of this distribution.

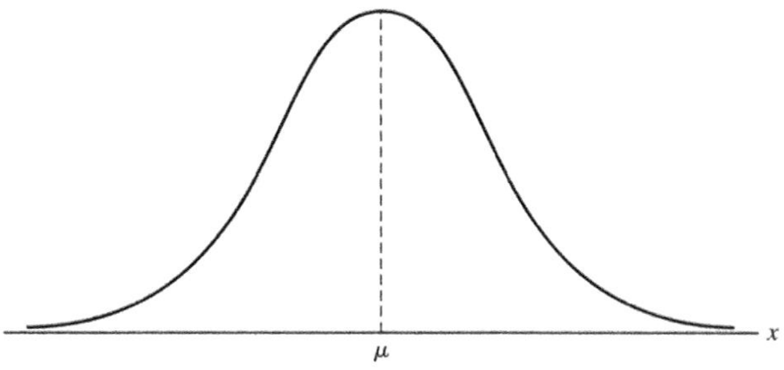

Fig. 3.1 Graph of normal distribution

Definition 3.35 Normal Distribution. A random variable X has a normal distribution with expectation μ and variance σ^2, *i. e.* , $X \sim N(\mu, \sigma^2)$, if and only if its probability density is given by

$$f(x) = \frac{1}{\sigma\sqrt{2\pi}} e^{-\frac{1}{2}\left(\frac{x-\mu}{\sigma}\right)^2}, \quad \text{for} \quad x \in \mathbb{R},$$

where $\sigma > 0$.

The graph of a normal distribution is shown in Fig. 3.1. It is shaped like the cross-section of a bell. μ and σ are the two parameters that play a key role in the shape of normal distribution.

Definition 3.36 Standard Normal Distribution. The normal random variable with $\mu = 0$ and $\sigma = 1$ is referred to as the standard normal random variable.

Theorem 3.28 If *X* has a normal distribution with the mean μ and the standard deviation σ, then $Z = \frac{X-\mu}{\sigma}$ has a standard normal distribution.

Proof See Freund et al. (2004).

Example 3.7 If X is a normal random variable with $\mu = 3$ and $\sigma = 4$, find $P(4 \leq X \leq 8)$.

Solution:

$$\begin{aligned} P(4 \leq X \leq 8) &= P\left(\frac{4-3}{4} \leq \frac{X-3}{4} \leq \frac{8-3}{4}\right) \\ &= P(0.25 \leq Z \leq 1.25) \\ &= P(Z \leq 1.25) - P(X \leq 0.25) \\ &= 0.8944 - 0.5987 \\ &= 0.2957. \end{aligned}$$

Definition 3.37 Multivariate Normal Random Variable. Let $X = (X_1, \ldots, X_k)$ be a *k*-dimensional random variable. *X* has a multivariate normal distribution with mean

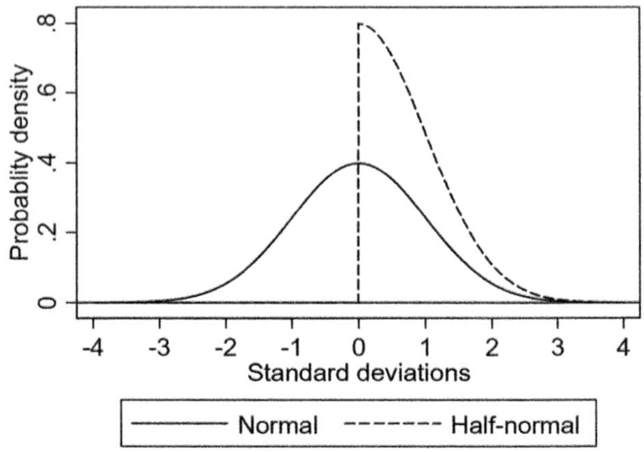

Fig. 3.2 PDFs of the normal distribution (mean zero) and half-normal distribution

μ and matrix variance–covariance Σ, *i.e.*, $X \sim N(\mu, \Sigma)$, if and only if its probability density is represented by

$$f_X(x_1, \ldots, x_k) = \frac{1}{\sigma\sqrt{(2\pi)^k \mid \Sigma \mid}} e^{-\frac{1}{2}(X-\mu)^T\Sigma^{-1}(X-\mu)}, \text{for } x_i \in \mathbb{R}, i = 1, \ldots, k.$$

Note that $\mu = \mathbb{E}(X) = [\mathbb{E}(X_1), \mathbb{E}(X_2), \ldots, \mathbb{E}(X_k)]$ and $\Sigma = [Cov(X_i, X_j); 1 \le i, j \le k]$.

Example 3.8 Suppose $Z = (X, Y)$ is a bivariate normal random variable. Let the mean vector and variance–covariance matrix of Z be $\mu = (\mu_X, \mu_Y)$ and $\Sigma = \begin{pmatrix} \sigma_X^2 & \rho\sigma_X\sigma_Y \\ \rho\sigma_X\sigma_Y & \sigma_Y^2 \end{pmatrix}$, respectively. Then, $f_{X,Y}(x, y) = \frac{1}{2\pi\sigma_X\sigma_Y\sqrt{1-\rho^2}} \exp\left(-\frac{1}{2(1-\rho^2)}\left[\frac{(x-\mu_X)^2}{\sigma_X^2} + \frac{(y-\mu_Y)^2}{\sigma_Y^2} - \frac{2\rho(x-\mu_X)(y-\mu_Y)}{\sigma_X\sigma_Y}\right]\right)$.

Note that ρ is the correlation coefficient between X and Y.

Definition 3.38 Half-normal distribution. A random variable *X* has a **half-normal distribution** and it is referred to as a half-normal random variable if and only if its probability density function is represented by $f(x; \sigma) = \frac{\sqrt{2}}{\sigma\sqrt{\pi}} \exp\left(-\frac{x^2}{2\sigma^2}\right), x > 0$.

Let Z follow a normal distribution, *i.e.* $Z \sim N(0, \sigma^2)$. Then, $X = |Z|$ follows a half-normal distribution. The graph of a normal distribution along with a half-normal distribution is shown in Fig. 3.2.

3.7 The Chi-Square Distribution

In probability theory and statistics, the chi-square distribution (χ^2-distribution) with k degrees of freedom is the distribution of the sum of the squares of k independent standard normal random variables. Due to the importance of this distribution, we will discuss it here.

Definition 3.39 Chi-Square Distribution. A random variable X has a chi-square distribution, *i.e.*, $X \sim \chi^2(k)$, and it is referred to as a chi-square random variable if and only if its probability density function is represented by

$$f(x,n) = \begin{cases} \dfrac{1}{2^{\frac{k}{2}}\Gamma\left(\dfrac{k}{2}\right)} x^{\frac{k-2}{2}} e^{-\frac{x}{2}}, & \text{for} \quad x > 0; \\ 0 \quad , & \text{otherwise;} \end{cases}$$

where k is referred to as the degrees of freedom.

Corollary 3.4 The mean and variance of the chi-square distribution are $\mu = \nu$ and $\sigma^2 = 2\nu$, respectively.

Theorem 3.29 Suppose $X_1, X_2, \ldots, X_n$ are n independent random variables having standard normal distributions. Then, $Y = \sum_{i=1}^{n} X_i^2$ has a chi-square distribution with n degrees of freedom.

Proof See Freund et al. (2004).

Theorem 3.30 Suppose that X is a k-dimension random vector, $X \sim N(\mu, \Sigma)$, where Σ is positive definite. Then $(X - \mu)^T\Sigma^{-1}(X - \mu)$ follows a chi-square distribution on k degrees of freedom.

Proof See Flury (2013).

Remark 3.2 The curves of the form $(X - \mu)^T\Sigma^{-1}(X - \mu) = \text{constant} > 0$ are ellipses or ellipsoids in higher dimensions.

From Example 3.8, to find an ellipse within X, which falls with probability α, we need to set $(X - \mu)^T\Sigma^{-1}(X - \mu)$ to be equal to the α-quantile of the chi-square distribution on two degrees of freedom. If U is a chi-square on two degrees of freedom, then its distribution function is represented by

$$F(u) = P(U \leq u) = \begin{cases} 0 \quad , & u \leq 0; \\ 1 - e^{-\frac{1}{2}u}, & u > 0. \end{cases}$$

And, therefore, the α-quantile is computed as follows:

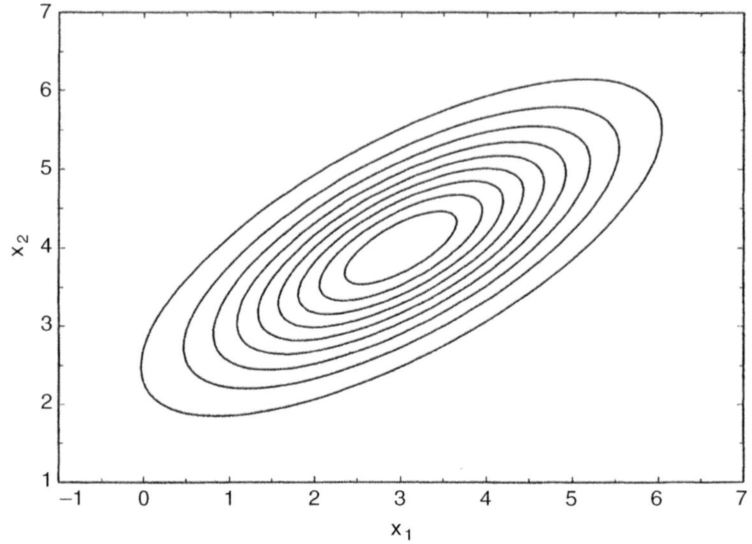

Fig. 3.3 Ellipses of the constant density of a bivariate normal distribution. The ellipses represent the regions within which X falls with probability $\alpha = 0.1, 0.2, \ldots, 0.9$

$$F(u) = \alpha \Rightarrow 1 - e^{-\frac{1}{2}u} = \alpha$$
$$\Rightarrow u = -2\log(1 - \alpha).$$

By choosing different values for $\alpha \in (0, 1)$, we will obtain different values for u. Therefore, for each α, we have a quadratic curve in the (x_1, x_2)-plane.

For instance, for $\mu = \begin{pmatrix} 3 \\ 4 \end{pmatrix}$ and $\Sigma = \begin{bmatrix} 2 & 1 \\ 1 & 1 \end{bmatrix}$, we have

$$(X - \mu)^T \Sigma^{-1}(X - \mu) = (x_1 - 3, x_2 - 4)\begin{bmatrix} 1 & -1 \\ -1 & 2 \end{bmatrix}\begin{pmatrix} x_1 - 3 \\ x_2 - 4 \end{pmatrix} = c^2,$$

where $c^2 = -2\log(1 - \alpha)$.

Figure 3.3 shows various curves for $\alpha = 0.1, 0.2, \ldots, 0.9$.

3.8 Sampling Distributions

As we know, the events of a statistical experiment are determined numerically by the random variables. The total set of observations surveyed is called **population** and the number of population members is called the **size of the population**. The observations are values of a random variable, and since each random variable has a probability distribution, each statistical population can be assigned a random variable and thus has a probability distribution. For example, if a random variable corresponding to the observations of a population is a normal random variable, that population is called the normal population.

Since we define a population as a set, it can have a finite or infinite number of members. On the other hand, since we do not have access to all observations in an infinite population, we do not know the distribution of the population, and we do not know its mean and variance. Therefore, we call the mean and variance of the population as population parameters that must be estimated. For this purpose, we need a sample of the population.

Definition 3.40 Random Sample. If $X_1, X_2, \ldots, X_n$ are independent and identically distributed random variables, we say that they constitute a random sample from the infinite population given by their same distribution.

Note: If $X_1, X_2, \ldots, X_n$ are n independent and identically distributed random variables with the same probability function, then the probability distribution of this random sample is $f(x_1, x_2, \ldots, x_n) = f(x_1)f(x_2)\ldots f(x_n)$.

Now, to estimate the population parameter, we first introduce the mean and variance of the sample. These calculated values from a random sample are called statistics and since these values depend on the sample and there are many random samples in the population, then each statistic is a random variable.

Definition 3.41 Statistic. Any function of random sample members that does not contain unknown parameters is called a statistic.

Definition 3.42 Sample mean and sample variance. If $X_1, X_2, \ldots, X_n$ constitute a random sample, then the sample mean is given by $\overline{X} = \frac{\sum_{i=1}^{n} X_i}{n}$, and the sample variance is given by $S^2 = \frac{\sum_{i=1}^{n} (X_i - \overline{X})^2}{n-1}$.

Note: $\overline{X}$ and S^2 are the two statistics.

Theorem 3.31 If $X_1, X_2, \ldots, X_n$ constitute a random sample from an infinite population with the mean μ and the variance σ^2, then $\mathbb{E}(\overline{X}) = \mu$ and $\operatorname{var}(\overline{X}) = \frac{\sigma^2}{n}$.

Theorem 3.32 If $X_1, X_2, \ldots, X_n$ constitute a random sample from an infinite population with the mean μ and the variance σ^2, then $\mathbb{E}(S^2) = \sigma^2$.

3.8.1 *Limit Theorems*

Limit theorems are the most important theoretical results in probability theory. These theorems are regarded as the results giving convergence of sequences of random variables or their distribution functions. Since random variables are functions with random influences, then different modes of convergence are involved in a sequence of random variables. The central limit theorem and the law of large numbers are the most important limit theorems, which we will be introducing in this section.

Theorem 3.33 The law of large numbers. If $X_1, X_2, \ldots, X_n$ constitute a random sample from an infinite population with the mean μ and variance σ^2, then for any $\epsilon > 0$, $\mathbb{P}(|\overline{X} - \mu| \geq \epsilon) \to 0$ as $n \to \infty$.

The law of large numbers states that the sample average converges in probability toward the expected value. In fact, as the sample size increases, the sample mean gets closer to the population mean.

Theorem 3.34 The central limit theorem. Let $X_1, X_2, \ldots, X_n$ constitute a random sample from an infinite population with the mean μ and variance σ^2. Then the distribution of $Z = \frac{\overline{X}-n\mu}{\sigma\sqrt{n}}$ as $n \to \infty$ is the standard normal distribution. That is, for $-\infty < a < \infty$,

$$\mathbb{P}\left(\frac{\overline{X} - n\mu}{\sigma\sqrt{n}} \leq a\right) \to \frac{1}{\sqrt{2\pi}} \int_{-\infty}^{a} e^{-x^2/2} dx \quad \text{as } n \to \infty.$$

3.9 Estimation Theory

In this section, we want to analyze and estimate population parameters using the statistics presented in the previous section. It is possible to examine population parameters in several ways. The most common of these is the classical estimation method, which provides the population parameters directly from a sample of the population.

The parameters of a population can be estimated in two ways; point estimation and interval estimation. Generally, the parameter of the population and the statistics that will be used to examine this parameter are denoted by θ and $\widehat{\Theta}$, respectively. The statistic used for the point estimation is called the estimator. Since one parameter may have multiple estimators, we need to know which estimator is better.

Definition 3.43 Unbiased estimator. A statistic $\widehat{\Theta}$ is an unbiased estimator of the parameter θ of a given distribution if and only if $\mathbb{E}\left(\widehat{\Theta}\right) = \theta$ for all possible values of θ.

Note: $\widehat{\Theta}$ is called a biased estimator if $\mathbb{E}\left(\widehat{\Theta}\right) \neq \theta$. The bias is then defined as the difference between $\mathbb{E}\left(\widehat{\Theta}\right)$ and θ, *i.e.* bias $= \mathbb{E}\left(\widehat{\Theta}\right) - \theta$.

Each parameter may have several unbiased estimators. If we must choose one of them, we usually take the one whose sampling distribution has the smallest variance.

Example 3.9 If $X_1, X_2, \ldots, X_n$ has a Bernoulli distribution with success parameter p, then the statistic $\overline{X} = \frac{\sum_{i=1}^{n} X_i}{n}$ is an unbiased estimator of p.

$$\text{Solution. } \mathbb{E}(\overline{X}) = \mathbb{E}\left(\frac{\sum_{i=1}^{n} X_i}{n}\right) = \frac{1}{n}\sum_{i=1}^{n} \mathbb{E}(X_i) = \frac{1}{n}np = p.$$

Definition 3.44 Minimum variance unbiased estimator. The estimator for the parameter θ of a given distribution that has the smallest variance of all unbiased estimators for θ is called the minimum variance unbiased estimator, or the best-unbiased estimator for θ.

Theorem 3.35 Let $\widehat{\Theta}$ be an unbiased estimator of θ and

$$\text{var}\left(\widehat{\Theta}\right) = \frac{1}{n.\mathbb{E}\left[\left(\frac{\partial \ln f(X)}{\partial \theta}\right)^2\right]},$$

then $\widehat{\Theta}$ is the unbiased estimator of θ with minimum variance.

The quantity in the denominator is referred to as the **information** about θ. Thus, the smaller the variance is, the greater the information.

Note: If $\widehat{\Theta}_1$ and $\widehat{\Theta}_2$ are two unbiased estimators of θ, where the variance of $\widehat{\Theta}_1$ is smaller than the variance of $\widehat{\Theta}_2$, then we say $\widehat{\Theta}_1$ is **relatively more efficient** than $\widehat{\Theta}_2$. The efficiency of $\widehat{\Theta}_1$ relative to $\widehat{\Theta}_2$, denoted eff $\left(\widehat{\Theta}_1, \widehat{\Theta}_2\right)$ is defined to be the ratio:

$$\text{eff}\left(\widehat{\Theta}_1, \widehat{\Theta}_2\right) = \frac{var\left(\widehat{\Theta}_2\right)}{var\left(\widehat{\Theta}_1\right)}.$$

Definition 3.45 The *mean square error* of a point estimator $\widehat{\Theta}$ is $MSE\left(\widehat{\Theta}\right) = \mathbb{E}\left[\left(\widehat{\Theta} - \theta\right)^2\right]$.

$MSE\left(\widehat{\Theta}\right)$ is also called the risk function of an estimator.

Definition 3.46 Consistent estimator. The statistic $\widehat{\Theta}$ is a consistent estimator of the parameter θ of a given distribution if and only if for each $c > 0$, $\lim_{n\to\infty} \mathbb{P}\left(\left|\widehat{\Theta} - \theta\right| \le c\right) = 1$ or, equivalently, $\lim_{n\to\infty} \mathbb{P}\left(\left|\widehat{\Theta} - \theta\right| > c\right) = 0$.

The previous definition says that when the size of the random sample is sufficiently large, we can be practically certain that the error made with a consistent estimator will be less than any small pre-assigned positive constant.

Theorem 3.36 If $\widehat{\Theta}$ is an unbiased estimator of the parameter θ and $var\left(\widehat{\Theta}\right) \to 0$ as $n \to \infty$, then $\widehat{\Theta}$ is a consistent estimator of θ.

Definition 3.47 Sufficient estimator. The statistic $\widehat{\Theta}$ is a **sufficient estimator** of the parameter θ of a given distribution if and only if for each value of $\widehat{\Theta}$, the conditional probability distribution or density of the random sample $X_1, X_2, \ldots, X_n$, given $\widehat{\Theta} = \theta$, is independent of θ.

The statistic $\widehat{\Theta}$ of a parameter θ which gives as much information about θ as is possible from the sample is called a sufficient estimator.

Definition 3.48 Minimal sufficient statistic. The sufficient statistic $\widehat{\Theta}$ is called the minimal sufficient if, for any other sufficient statistic such as $\widehat{\widehat{\Theta}}$ and the arbitrary function *f*, we have $\widehat{\Theta} = \mathrm{f}\left(\widehat{\widehat{\Theta}}\right)$.

It should be noted that the sufficient statistic is not unique. By Definition 3.48, a sufficient statistic, which is a function of all-sufficient statistics, is called a Minimal Sufficient. Thus, the minimal sufficient statistics can be considered as the most effective sufficient statistics for the parameter θ, which is simpler than all the sufficient statistics.

Theorem 3.37 The Rao–Blackwell theorem. Let $\widehat{\theta}$ be an unbiased estimator for θ such that $\mathrm{Var}\left(\widehat{\theta}\right) < \infty$. If $\widehat{\Theta}$ is a sufficient statistic for θ, define $\widehat{\theta}^* = \mathbb{E}\left(\widehat{\theta}|\widehat{\Theta}\right)$. Then, for all θ, $\mathbb{E}\left(\widehat{\theta}^*\right) = \theta$ and $var\left(\widehat{\theta}^*\right) \leq var\left(\widehat{\theta}\right)$.

The Rao–Blackwell theorem says that, if $\widehat{\theta}$ is an unbiased estimator for θ and if $\widehat{\Theta}$ is a sufficient statistic for θ, then there is a function of $\widehat{\Theta}$ that is also an unbiased estimator for θ and has variance no larger than $\widehat{\theta}$. This theorem can be used to find an unbiased estimator with the least variance (*Minimum Variance Unbiased Estimator, MVUE*). To find it, it is enough to find the uncertainty estimator and obtain its conditional expectation in terms of sufficient parameter statistics. The resulting estimator will have a smaller variance than the variance of the initial estimator. To determine the best estimator in the class or set of unbiased estimators, the Lehmann–Scheffe theorem should be used. This theorem ensures that the unbiased estimator created by the Rao–Blackwell theorem has less variance than any other unbiased estimator.

Definition 3.49 The statistic $\widehat{\Theta}$ for its distribution family is complete if the following relation exists for any measurable function *g* and any parameter value θ. That is,

$$\mathbb{E}\left(g\left(\widehat{\Theta}\right)\right) = 0 \rightarrow \mathbb{P}\left(g\left(\widehat{\Theta}\right)\right) = 0) = 1.$$

In the sufficient statistic, we consider a statistic that has the most information about the unknown parameter. We also know that a sufficient statistic that is a function of all-sufficient statistics is called a minimal sufficient statistic and it is best to use this statistic for estimation. But the goal of creating a complete statistic is to produce or select a statistic that can provide us with the smallest amount of information about the parameter. This means that in a minimal sufficient statistic, there may be additional information that is not relevant to draw inference about the population parameter. Selecting "Complete Minimal Sufficient Statistic" will result in a statistic that stores information only about the population parameter and has no superfluous information.

Theorem 3.38 Lehmann–Scheffe. Let $\widehat{\Theta}$ be a complete sufficient statistic. If there are unbiased estimators, then there exists a unique MVUE. We can obtain the MVUE as $\widehat{\widehat{\Theta}} = \mathbb{E}\left(\widehat{\theta}|\widehat{\Theta}\right)$, for any unbiased $\widehat{\theta}$. The MVUE can also be characterized as the unique unbiased function $\widehat{\widehat{\Theta}} = \varphi\left(\widehat{\Theta}\right)$ of the complete sufficient statistic $\widehat{\Theta}$.

This theorem helps to identify the best estimator from the class of unbiased estimators. This is a complement to Rao–Blackwell's theorem. Using this theorem, we can show under what conditions the unbiased estimator with the least variance is unique. This way, a "Uniformly Minimum Variance Unbiased Estimator-UMVUE" estimator can be obtained. Uniform means that this estimator has the least variance in the class of unbiased estimators for all points in the parametric space.

3.9.1 *The Method of Maximum Likelihood*

One of the most popular methods for estimating parameters is the method of maximum likelihood. The advantages of this method are that it yields sufficient estimators and the maximum likelihood estimators are the minimum variance unbiased estimators.

Definition 3.50 Maximum likelihood estimator (MLE). Let $x_1, x_2, \ldots, x_n$ be the values of a random sample from a population with the parameter θ. The likelihood function of the sample is then represented by $L(\theta) = f(x_1, x_2, \ldots, x_n|\theta) = \prod_{i=1}^{n} f(x_i|\theta)$ for values of θ within a given domain.

Note that $f(x_1, x_2, \ldots, x_n; \theta)$ is the value of the joint probability distribution or joint density of random variables $X_1, X_2, \ldots, X_n$ at $X_1 = x_1, X_2 = x_2, \ldots, X_n = x_n$. We refer to the value of θ that maximizes $L(\theta)$ as the maximum likelihood estimator of θ. It is usually customary and easier to maximize the logarithm of the likelihood, $\ln L(\theta)$. To maximize $ln\ L(\theta)$, we take the derivative of $ln\ L(\theta)$ concerning θ and set the expression equal to 0, *i.e.* $\frac{\partial \ln L(\theta)}{\partial \theta} = 0$.

Example 3.10 If $x_1, x_2, \ldots, x_n$ are the values of a random sample from an exponential population, find the maximum likelihood estimator of its parameter θ.

Solution According to the definition of likelihood function, we have $L(\theta) = \prod_{i=1}^{n} f(x_i|\theta) = \left(\frac{1}{\theta}\right)^n . e^{-\frac{1}{\theta}\left(\sum_{i=1}^{n} x_i\right)}$. Differentiating $lnL(\theta)$ with respect to θ yields $\frac{d\ \ln L(\theta)}{d\theta} = -\frac{n}{\theta} + \frac{1}{\theta^2} \cdot \sum_{i=1}^{n} x_i = 0$. By solving this equation, we get the maximum likelihood estimate as $\theta = \frac{1}{n} \sum_{i=1}^{n} x_i = \overline{x}$. Hence, the maximum likelihood estimator is $\widehat{\Theta} = \overline{X}$.

3.9.2 *Linear Regression Model*

One of the popular methods for studying the causal relationship between independent and dependent variables is the linear regression method. There are two types of relationships between variables; *deterministic* and *probabilistic*. In the deterministic form, the relationship between the two variables is exact. For example, we might have $Y = \beta X$,where the value of Y is determined by X. On the other hand, in the probabilistic form, the relationship between variables involves random components or random error. For example, we might have $Y = \beta X + \epsilon$ containing two components; a deterministic component βX plus a random error ϵ.

3.9.3 *General Linear Model*

Let the model for linear regression be represented by

$$Y = \beta X + \epsilon \text{ with } \epsilon \sim N\left(0, \sigma^2 I\right),$$

where

$$Y = \begin{bmatrix} Y_1 \\ Y_2 \\ \vdots \\ Y_n \end{bmatrix}, \quad X = \begin{bmatrix} 1 & x_{12} & \cdots & x_{1p} \\ 1 & x_{22} & \cdots & x_{2p} \\ \vdots & \vdots & \ddots & \vdots \\ 1 & x_{n2} & \cdots & x_{np} \end{bmatrix}, \quad \beta = \begin{bmatrix} \beta_0 \\ \beta_1 \\ \vdots \\ \beta_p \end{bmatrix} \quad \text{and } \epsilon = \begin{bmatrix} \epsilon_1 \\ \epsilon_2 \\ \vdots \\ \epsilon_n \end{bmatrix}.$$

Note that $\mathbb{E}(\epsilon_i \epsilon_i) = 0, i \neq j$ and $\mathbb{E}(\epsilon_i \epsilon_i) = \sigma^2, \quad i = j.$

Our problem is to choose an estimation of linear regression of the form $Y = \widehat{\beta}X + \boldsymbol{e}$, where $\widehat{\beta}$ is a $(p + 1)$ column vector as an estimation of the vector β and $\boldsymbol{e}$ is an n column vector of residuals.

3.9.4 Ordinary Least Square Method (OLS)

The problem of classical linear model estimation requires estimation of the unknown parameters $\beta_0, \beta_1, \ldots, \beta_n$ and σ^2. The ordinary least squares method selects $\beta_0, \beta_1, \ldots, \beta_n$ values to minimize the sum of squares of the errors, *i.e.* $S = \sum_{i=1}^{n} \left(Y_i - \beta_o - \beta_1 x_{i1} - \ldots - \beta_p x_{ip}\right)^2$. This can also be represented as

$$\begin{aligned} S &= (Y - \beta X)^T (Y - \beta X) \\ &= Y^T Y - \beta^T X^T Y - Y^T X \beta + \beta^T X^T X \beta \\ &= Y^T Y - 2\beta^T X^T Y + \beta^T X^T X \beta. \end{aligned}$$

Now, we must derive S in terms of β and then set it to zero. Differentiating S with respect to β and equating it to zero yields

$$\frac{\partial S}{\partial \beta} = -2X^T Y + 2X^T X \widehat{\beta} = 0.$$

Solving this equation yields $\widehat{\beta} = \left(X^T X\right)^{-1} X^T Y$.

Note that since $\frac{\partial^2 S}{\partial \beta^2} = 2X^T X$ is a positive definite matrix, then $\widehat{\beta}$ will minimize S.

Definition 3.51 The Best Linear Unbiased Estimator (BLUE) of a parameter β based on data X

1. is a linear function of X, *i.e.* the estimator can be written as $\widehat{\beta}X$;
2. is unbiased, *i.e.* $\mathbb{E}\left(\widehat{\beta}X\right) = \beta$; and
3. has the smallest variance among all the unbiased linear estimators.

Finally, we end this chapter by presenting an important theorem below.

Theorem 3.39 Gauss–Markov Theorem. If $\widehat{\beta}$ is the ordinary least square estimator of β in the classical linear regression model, and if $\widehat{\widehat{\beta}}$ is any other linear unbiased estimator of β, then $\operatorname{var}\left(c'\widehat{\widehat{\beta}}\right) \geq \operatorname{var}\left(c'\widehat{\beta}\right)$, where c is any constant vector of the appropriate order.

In statistics, the Gauss–Markov theorem states that in a linear model whose errors have zero expectations are uncorrelated and have equal variances, the BLUE for the system coefficients is the least squares estimator.

References

Flury, B. (2013). *A first course in multivariate statistics*. Springer Science & Business Media.

Freund, J. E., Miller, I., & Miller, M. (2004). *John E. Freund's Mathematical statistics: With applications*. Pearson Education India.

Chapter 4
Stochastic Data Envelopment Analysis

In traditional DEA models, the technologies are developed using the premise that inputs and outputs are precisely measured and are, therefore, deterministic. However, in practical situations, the general production processes are often stochastic. The stochastic production relationship in a DEA setting may arise in different situations, for example, when stochastic variations in inputs and outputs affect the production frontier; when inputs and outputs are faced with stochastic prices while measuring allocative efficiency; when the slacks obtained from the DEA efficiency frontier are analyzed in terms of their statistical distribution; when an economic method is applied to estimate the stochastic production frontier; etc. (Sengupta, 1990). Over the last two decades, many researchers have proposed DEA-based models with stochastic data. Sengupta (2000) applied a stochastic DEA model using mathematical expectations for random inputs and outputs. Banker (1993) added statistical elements to DEA and developed an approach aimed at influencing statistical noise in inference. Many studies (*e.g.*, Cooper et al., 1996, 1998; Land et al., 1993; Olesen & Petersen, 1995, 2016) have introduced chance-constrained programming in DEA to accommodate random changes in data. Banker (1986) proposed a related semi-parametric stochastic frontier analysis (SFA) based on a minimization of the sum of the absolute value of all composed error terms. For parametric and semi-parametric models, see Banker (1989, 1996), Banker and Chang (1995), Banker et al. (1994, 2015), and Banker and Maindiratta (1992). Additional approaches and applications can be found in Charles and Cornillier (2017), Charles and Udhayakumar (2012), Charles et al. (2018), Grosskopf (1996), Horrace and Schmidt (1996), Simar (1996), Simar and Wilson (1998), and Udhayakumar et al. (2011), among others.

This chapter deals with the development of the stochastic DEA approach. In this approach, features such as "efficient" and "inefficient" are replaced with "probably efficient" and "probably inefficient." Furthermore, the "E-model" chance-constrained programming based on the mathematical expectation to obtain the expected value of the objective function and the "P-model" chance-constrained

A. Amirteimoori et al., *Stochastic Benchmarking*, International Series in Operations Research & Management Science 317,
https://doi.org/10.1007/978-3-030-89869-4_4

programming model based on the highest probability of occurrence of the objective function will be introduced.

4.1 Stochastic Technology Set

In deterministic DEA, the observed input/output data are used to construct an empirical technology set. An axiomatic foundation is needed to construct such a technology set. Under a stochastic environment, these axioms must be redefined to construct a stochastic technology set. Assuming x and y, respectively, as the vectors of inputs and outputs for a firm, the technology set (T) is defined as

$$T = \{(x, y) | y \text{ can be produced from } x\}.$$

For empirical construction of T from the set of n DMUs, we need the following axioms:

i) *Strong disposability*: $\forall (x, y) \in T;\ \ \forall x', y' \ \& \ x' \geq x, y' \leq y \Rightarrow (x', y') \in T.$
ii) *Ray unboundedness* (*CRS*): $(x, y) \in T \Rightarrow \lambda(x, y) \in T, \lambda \geq 0.$
iii) *Convexity*: $(x, y) \in T \ \ \& \ \ (x', y') \in T \Rightarrow \lambda(x, y) + (1 - \lambda)(x', y') \in T, 0 \leq \lambda \leq 1.$
iv) *Inclusion of observations*: $\forall j = 1, \ldots, n, (x_j, y_j) \in T.$
v) *Minimal extrapolation*: T is the smallest set satisfying the set of imposed maintained axioms.

For the construction of a stochastic technology set, we need all the above axioms except (iv). Axiom (iv) implies that if the observations are random, the probability of those observations being outside the technology set is zero. Olesen and Petersen (1995) proposed a relaxation on axiom (iv) that permits a stochastic variation around the frontier, *i.e.* it permits observations to fall on both sides of the frontier.

4.2 Chance-Constrained Empirical Technology Set

Let $\boldsymbol{w} \equiv (u^T, v^T)^T \in \mathbb{R}^{s+m}$, wherein $u \in \mathbb{R}^s$ and $v \in \mathbb{R}^m$ represent, respectively, the vector of output and input multipliers. Let the sample of input and output vectors be normally and jointly distributed, with a mean input vector $\overline{X}_j$, a mean output vector $\overline{Y}_j$, and a variance–covariance matrix Λ_j, $j = 1, \ldots, n$. Further, let the firm j^{th} output–input vector be denoted as $Z_j \equiv (Y_{1j}, \ldots, Y_{sj}, X_{1j}, \ldots, X_{mj}) \in \mathbb{R}^{s+m}$.

Let us define the following set:

$$V(\alpha) \equiv \left\{ \left(\boldsymbol{w}^T, \omega\right)^T \in \mathbb{R}^{s+m+1} \middle| \mathbb{P}\left[\left(Z_{1j}, \ldots, Z_{s+m,j}\right) \in H(\boldsymbol{w}, \omega)\right] = \alpha \right\}, \tag{4.1}$$

where

$$H(\boldsymbol{w}, \omega) \equiv \{z \in \mathbb{R}^{s+m} | w_1 z_1 + \ldots + w_{s+m} z_{s+m} + \omega \leq 0\}. \tag{4.2}$$

By using the assumption of joint normality, $V(\alpha)$ can be transformed into the following certainty-equivalent form:

$$\begin{aligned} \overline{V}(\alpha) \equiv \Big\{ \left(\boldsymbol{w}^T, \omega\right)^T \in \mathbb{R}^{s+m+1} \Big| & \left(\widehat{\overline{Y}}_{1j}, \ldots, \widehat{\overline{Y}}_{sj}, \widehat{\overline{X}}_{1j}, \ldots, \widehat{\overline{X}}_{mj}\right) \boldsymbol{w} + \omega \\ & + \Phi^{-1}(\alpha) \left(\boldsymbol{w}^T \widehat{\Lambda}_{\rm j}\, \mathbf{w}\right)^{\frac{1}{2}} \\ & = 0 \Big\}, \end{aligned} \tag{4.3}$$

where $\Phi(\cdot)$ is the standard normal distribution function and $\widehat{\overline{Z}}_j$ and $\widehat{\Lambda}_j$ are, respectively, the maximum likelihood estimates of $\overline{Z}_j$ and Λ_j, for all j. For a given $(\boldsymbol{w}^T, \omega)^T \in \overline{V}(\beta)$, this vector defines a half-space $H(\boldsymbol{w}, \omega)$ with a probability mass on this half-space equal to α. We define $C(\alpha)$ as the intersection of all such half-spaces from $(\boldsymbol{w}^T, \omega)^T \in \overline{V}(\alpha)$. In other words,

$$\begin{aligned} C(\alpha) \equiv & \bigcap_{(\boldsymbol{w}^T, \omega)^T \in \overline{V}(\alpha)} H(\boldsymbol{w}, \omega) \\ = & \{z \in \mathbb{R}^{s+m} | w_1 z_1 + \ldots + w_{s+m} z_{s+m} + \omega \leq 0, \forall (w, \omega) \in \overline{V}(\alpha)\}. \end{aligned} \tag{4.4}$$

Now, we define a joint confidence region for the $s + m$ dimensional normal distribution as

$$D(\gamma) = \left\{z \in \mathbb{R}^{s+m} |\, F(z_1, \ldots, z_{s+m}) < c^2, \mathbb{P}\left(\chi^2_{(s+m)} \leq c^2\right) = \gamma\right\}, \tag{4.5}$$

where

$$\begin{aligned} & F(z_1, \ldots, z_{s+m}) \equiv \\ & \left[\left[\left(z_1 - \overline{Z}_1\right), \ldots, \left(z_{s+m} - \overline{Z}_{s+m}\right)\right] \Lambda^{-1} \left[\left(z_1 - \overline{Z}_1\right), \ldots, \left(z_{s+m} - \overline{Z}_{s+m}\right)\right]\right], \end{aligned} \tag{4.6}$$

wherein $\chi^2_{(f)}$ is a random variable with a Chi-square distribution with f degrees of freedom. $D(\gamma)$ is the joint confidence region for $(Z_1, \ldots, Z_{s+m})^T$ on level γ and consists of the interior of one of the contour ellipsoids from the $s + m$ dimensional normal distribution. Note that $\mathbb{P}[(Z_1, \ldots, Z_{s+m}) \in D(\gamma)] = \gamma$.

Theorem 4.1 Let a sample of output–input vectors be distributed joint-normally with a mean vector $\overline{Z} \equiv \left[\overline{Y}_1, \ldots, \overline{Y}_s, \overline{X}_1, \ldots, \overline{X}_m\right] \in \mathbb{R}^{s+m}$ and an $(s + m) \times (s + m)$ variance–covariance matrix Λ of full rank. Also, let $H(\boldsymbol{w}, \omega)$, $C(\alpha)$, $D(\gamma)$, and $F(z_1, \ldots, z_{s+m})$ be defined as in (4.2), (4.4), (4.5), and (4.6). Then, $C(\alpha) = D(\gamma)$, where $\alpha = \Phi(c)$ and $\alpha = \mathbb{P}\left(\chi^2_{s+m} \leq c^2\right)$.

Proof See Olesen and Petersen (1995).

Corollary 4.1. Consider Z as the output–input vector with joint-normal distribution with mean vector $\overline{Z} \equiv \left[\overline{Y}_1, \ldots, \overline{Y}_s, \overline{X}_1, \ldots, \overline{X}_m\right] \in \mathbb{R}^{s+m}$, and a full rank $(s+m) \times (s+m)$ variance–covariance matrix, Σ. Also, let:

$$D(\gamma) = \left\{z \in \mathbb{R}^{s+m} \mid F(z_1, \ldots, z_{s+m}) < c^2, \mathbb{P}\left(\chi^2_{(s+m)} \le c^2\right) = \gamma\right\},$$

$$V(\beta) \equiv \left\{\mathbf{w}^T \in \mathbb{R}^{s+m+1} \mid \mathbb{P}\left[\left(Z_{1j}, \ldots, Z_{s+mj}\right) \in H(\mathbf{w})\right] = \beta\right\},$$

$$H(\mathbf{w}) \equiv \{z \in \mathbb{R}^{s+m} \mid w_1 z_1 + \ldots + w_{s+m} z_{s+m} \le 0\},$$

$$\overline{V}(\beta) \equiv \left\{\mathbf{w}^T \in \mathbb{R}^{s+m+1} \mid \left(\widehat{\overline{Y}}_{1j}, \ldots, \widehat{\overline{Y}}_{sj}, \widehat{\overline{X}}_{1j}, \ldots, \widehat{\overline{X}}_{mj}\right)\mathbf{w} + \Phi^{-1}(\beta)\left(\mathbf{w}^T \widehat{\Lambda}_j \mathbf{w}\right)^{\frac{1}{2}} = 0\right\},$$

$$C(\alpha) \equiv \bigcap_{\mathbf{w}^T \in \overline{V}(\alpha)} H(\mathbf{w}) = \{z \in \mathbb{R}^{s+m} \mid w_1 z_1 + \ldots + w_{s+m} z_{s+m} \le 0, \ \forall \mathbf{w} \in \overline{V}(\beta)\},$$

then, $C(\beta) = cone(D(\gamma))$, in which $\beta = \Phi(c)$ and $\alpha = \mathbb{P}\left(\chi^2_{s+m} \le c^2\right)$.

Proof See Olesen and Petersen (1995).

Now, to construct a technology set for firms with random inputs and outputs, we first define the inclusion of random observations as follows (with axiom iv' replacing the previous axiom iv):

iv') *Inclusion of observations within confidence regions*: The random observed input–output vectors are distributed by a probability law with a mean vector $\left(\overline{X}, \overline{Y}\right)$, and covariance matrix Σ. Only observations that fall within certain confidence regions are assumed to belong to the chance-constrained empirical technology set.

Consequently, by accepting the axioms (i), (ii), (iii), (v), and (iv') and by taking γ_j as the error level of the firm j, we define the stochastic technology set (T^s) as

$$T^s = \left\{\left(x^T, y^T\right)^T \in \mathbb{R}^{m+s}_{+} \mid \exists \left(\widehat{x}_j^T, \widehat{y}_j^T\right)^T \in D_j(1-\gamma_j) \text{ and } \lambda_j \ge 0, \ j \in \{1, \ldots, n\} \ \sum_{j=1}^{n} \lambda_j \widehat{x}_j \le x \text{ and } \sum_{j=1}^{n} \lambda_j \widehat{y}_j \ge y, \right\},$$

where $D_j(1-\gamma_j) =$

$$\left\{\left(x^T, y^T\right)^T \in \mathbb{R}^{m+s}_{+} \mid \left[\left(x-\overline{x}_j\right)^T, \left(y-\overline{y}_j\right)^T\right]\Sigma_j^{-1}\left[\left(x-\overline{x}_j\right)^T, \left(y-\overline{y}_j\right)^T\right]^T \le c_j^2\right\}.$$

Consider the vector $A = (\alpha_1, \ldots, \alpha_n)$, and let $\mathcal{P}(A)$, *i.e.*, the set of virtual multipliers under vector A, be defined as

Table 4.1 Mean vectors and covariance matrices of three DMUs (Olesen & Petersen, 1995)

	Mean			Covariance matrix Σ_j	
	$\overline{X}_j$	$\overline{Y}_j$		X_j	Y_j
DMU 1	2	5	X_1	0.404	−1.2
			Y_1	−1.2	3.604
DMU 2	4.5	5.8	X_2	0.488	0.264
			Y_2	0.264	0.148
DMU 3	0.9	3.3	X_3	0.068	0.048
			Y_3	0.048	0.040

$$\mathcal{P}(A) = \left\{ \left(u^T, -v^T\right)^T \in \mathbb{R}^{s+m} \middle| \, u \geq 0, v \geq 0, \mathbb{P}\left(u^T \widetilde{y}_j - v^T \widetilde{x}_j \leq 0\right) \geq 1 - \alpha_j \, (\forall j) \right\}. \tag{4.7}$$

By transforming the chance constraints (4.7) into their equivalent deterministic form, we get

$$\mathcal{P}(A) = \left\{ \left(u^T, -v^T\right)^T \in \mathbb{R}^{s+m} \middle| \; u \geq 0, v \geq 0, \; u^T y_j - v^T x_j + \Phi^{-1}\left(1 - \alpha_j\right) \sqrt{(u^T, -v^T)\Sigma(u^T, -v^T)^T} \leq 0 \right\}.$$

Since $\alpha \leq 0/5$, then $u^T y_j - v^T x_j + \Phi^{-1}\left(1 - \alpha_j\right) \sqrt{(u^T, -v^T)\Sigma(u^T, -v^T)^T}$ is a convex function, generally referred to as a convex cone.

Olesen and Petersen (1995), in Theorem 5, suggested that T^s is presented in terms of $\mathcal{P}(A)$ as

$$T^s = \left\{ \left(x^T, y^T\right)^T \in \mathbb{R}_+^{m+s} \, | \forall \left(u^T, -v^T\right)^T \in \mathcal{P}(A) : u^T - v^T x \leq 0 \right\}.$$

Finally, the production frontier for T^s can be presented as

$$E(T^s) = \left\{ \left(x^T, y^T\right)^T \in \mathbb{R}_+^{m+s} \, | \forall \left(u^T, -v^T\right)^T \in \mathcal{P}(A) : u^T - v^T x = 0 \right\}.$$

To empirically illustrate the T^s and $E(T^s)$, we consider an example from Olesen and Petersen (1995) where there are three DMUs with each one using one input to produce one output, and these units are observed over six consecutive periods. Suppose that j^{th} DMU consumes $X_{j,\,t}$ to produce $Y_{j,\,t}$ at time t, where $j = 1, 2, 3$ and $t = 1, \ldots, 6$. Moreover, suppose that $(X_{j,\,t}, Y_{j,\,t})$ are observations from three two-dimensional normal distributions.

The input–output mean vectors and covariance matrices of the estimated samples are shown in Table 4.1.

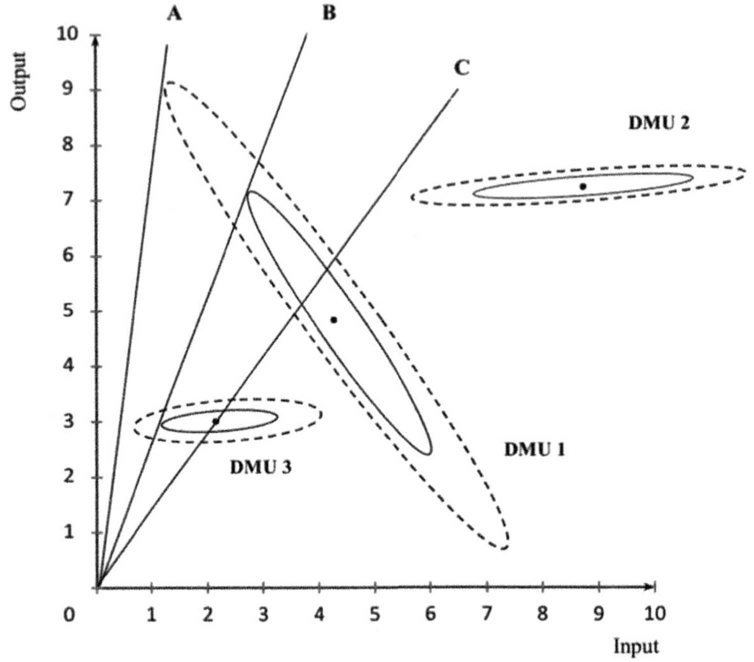

Fig. 4.1 The confidence regions for the three DMUs at probability levels $\gamma = 0.714$ and 0.8186

A portion of the input–output space consisting of three confidence zones $D_j(\gamma)$, at probability level γ, for DMU_j, $j = 1, 2, 3$, determined from the data, is shown in Fig. 4.1. Each confidence region is an ellipsoid, with the center and axes determined by the estimated mean and covariance. The algebraic expression for the confidence region $D_1(0.714)$ for DMU_1 at probability level 0.714 is as follows:

$$D_1(0.714) \equiv \left\{(y,x) \in \mathbb{R}^2 \mid \left(x - \overline{X}_1, y - \overline{Y}_1\right) \Sigma_j^{-1} \left(x - \overline{X}_1, y - \overline{Y}_1\right)^T \leq (1.5811)^2\right\}$$
$$= \left\{(y,x) \in \mathbb{R}^2 | \begin{bmatrix} x-2 \\ y-5 \end{bmatrix}^T \begin{bmatrix} 0.404 & -1.2 \\ -1.2 & 3.604 \end{bmatrix}^{-1} \begin{bmatrix} x-2 \\ y-5 \end{bmatrix} \leq (1.5811)^2 \right\},$$

where $\mathbb{P}\left(\chi^2_{(2)} \leq (1.5811)^2\right) \approx 0.714$ and $\chi^2_{(f)}$ have a chi-square random variable with f degrees of freedom.

As mentioned earlier, the probability level of the chance constraint for the confidence region $D_1(0.714)$, as will be shown below, is obtained from the relation $\Omega = \Phi(1.5811) \approx 0.9431$. Hence, for example, a set of input–output vectors in cone C, generated by $D_1(0.714)$, as a subset of cone $\overline{C}$, generated by the equivalent deterministic form of chance constraints in E at probability level 0.9431, can be expressed as follows:

$$C \equiv cone(D_1(0.714)) = \{(y,x) \in \mathbb{R}^2 | (y,x) = tD_1(0.714), t \geq 0\},$$

$$E \equiv \{(y,x) \in \mathbb{R}^2_+ | uy - vx \leq 0, \ \ \forall u, v \geq 0, \mathbb{P}(\mathrm{uY_1} - \mathrm{vX_1} \leq 0) \geq 0.9431\},$$

$$\overline{E} \equiv \{(y,x) \in \mathbb{R}^2_+ | \ uy - vx \leq 0, \forall u, v \geq 0, u\,\overline{Y}_1 - v\,\overline{X}_1 + \Phi^{-1}(0.9431)\sqrt{[u,v]\Sigma_1[u,v]^T} \leq 0\},$$

wherein $\mathrm{C} \subseteq \overline{\mathrm{E}}$. By combining the strong disposability axiom of inputs and outputs and the generated cone by $D_1(0.714)$, $\overline{E}$ can alternatively be expressed, based on the confidence region $D_1(0.714)$, as

$$\overline{E} = Cone(D_1(0.714)) + \mathbb{R}_- \times \mathbb{R}_+ = \{(x,y) \in \mathbb{R}^2 | y \leq ty', x \geq tx', (y',x') \in D_1(0.714), \ \ t \geq 0\}.$$

In Fig. 4.1, line OA from origin to A is the best practice frontier generated by chance constraints at the probability level 0.9431, *i.e.* it is the frontier generated by an envelopment of the confidence region at probability level 0.714. As it can be seen, DMU_1 is the only efficient unit at probability level 0.9431. Line OB is the best practice frontier at probability level 0.8186, where $\Phi^{-1}(0.8186) \approx 0.91$. This frontier is a supporting hyperplane for both confidence regions $D_1(0.339)$ and $D_3(0.339)$. Therefore, since the frontier supports both confidence regions, then DMU_1 and DMU_3 are efficient at probability level 0.8186. Finally, line OC is the best practice frontier at probability level 0.5, *i.e.* the frontier corresponding to the traditional DEA CCR model of mean input–output vectors. Note that the points located in the center of each ellipsoid in Fig. 4.1 are the mean input–output vector of each DMU.

4.3 Stochastic Data Envelopment Analysis Models

The efficiency evaluation of firms is one of the key issues from the viewpoint of managers. Knowledge of this issue at present will be a useful tool for adopting important and appropriate strategies to increase efficiency in the future. Stochastic programming is one of the applied tools for such forecasting purposes. By performing random sampling of the data from each firm over a specific period, by investigating the behavior of these data, and by using probability distribution tests, we estimate the efficiency in the future. As a result, the efficiency of each unit has a stochastic nature. In chance-constrained DEA models, the probability of failing to achieve the expected value is equal to α. The obtained results are true with 100 $(1 - \alpha)\%$ confidence.

4.3.1 Chance-Constrained DEA Model (E-Model)

The formal "E-model" of chance-constrained DEA under the CRS specification of technology (Land et al., 1993) is set up as

$$\theta_o^* = \min \theta \tag{4.8}$$
$$\mathbb{P}\left\{\sum\nolimits_{j=1}^{n} \lambda_j \widetilde{x}_{ij} \le \theta \widetilde{x}_{io}\right\} \ge 1-\alpha \ (\forall i),$$
$$\mathbb{P}\left\{\sum\nolimits_{j=1}^{n} \lambda_j \widetilde{y}_{rj} \ge \widetilde{y}_{ro}\right\} \ge 1-\alpha \ (\forall r),$$
$$\lambda_j \ge 0 \ (\forall j),$$

where $\alpha \in [0,1]$ is a user-defined parameter. By considering random inputs and outputs and estimating their probability distributions, we seek to estimate the efficiency in the future.

To evaluate each firm, model (4.8) should be converted into a deterministic form. Toward this end, consider the chance constraint of the ith input: $\mathbb{P}\left\{\sum\nolimits_{j=1}^{n} \lambda_j \widetilde{x}_{ij} - \theta \widetilde{x}_{io} \le 0\right\} \ge 1-\alpha$. By introducing an "external slack," $\eta_i \ge 0$, this inequality constraint can be transformed into an equality constraint: $\mathbb{P}\left\{\sum\nolimits_{j=1}^{n} \lambda_j \widetilde{x}_{ij} - \theta \widetilde{x}_{io} \le 0\right\} = 1-\alpha+\eta_i$.

Remark 4.1. Let X be a random variable and a, b, c, and d be constants. If $a \ge b$ and $\mathbb{P}(X \le a) = c$, then $\mathbb{P}(X \le b) = d$, where $d \le c$.

Considering this remark and defining a slack variable $s_i^- \ge 0$, we have

$$\mathbb{P}\left\{\sum\nolimits_{j=1}^{n} \lambda_j \widetilde{x}_{ij} - \theta \widetilde{x}_{io} \le -s_i^-\right\} = 1-\alpha. \tag{4.9}$$

Let $\widetilde{h}_i = \sum\nolimits_{j=1}^{n} \lambda_j \widetilde{x}_{ij} - \theta \widetilde{x}_{io}$ and considering the fact that any random combination of normal random variables has a normal distribution, we have $\widetilde{h}_i \sim N\left(h_i, \left(\sigma_i^I(\lambda,\theta)\right)^2\right)$, where $h_i = \mathbb{E}\left[\widetilde{h}_i\right] = \mathbb{E}\left[\sum\nolimits_{j=1}^{n} \lambda_j \widetilde{x}_{ij} - \theta \widetilde{x}_{io}\right] = \sum\nolimits_{j=1}^{n} \lambda_j x_{ij} - \theta x_{io}$ and $\left(\sigma_i^I(\lambda,\theta)\right)^2 = var\left[\widetilde{h}_i\right] = var\left[\sum\nolimits_{j=1}^{n} \lambda_j \widetilde{x}_{ij} - \theta \widetilde{x}_{io}\right] = var\left[\sum\nolimits_{j=1}^{n} \lambda_j \widetilde{x}_{ij}\right] + var[\theta \widetilde{x}_{io}] - 2cov\left[\sum\nolimits_{j=1}^{n} \lambda_j \widetilde{x}_{ij} - \theta \widetilde{x}_{io}\right] = \sum\limits_{j=1}^{n} \sum\limits_{k=1}^{n} \lambda_j \lambda_k cov\left(\widetilde{x}_{ij}, \widetilde{x}_{ik}\right) + \theta^2 var(\widetilde{x}_{io}) - 2\theta \sum\limits_{j=1}^{n} \lambda_j cov\left(\widetilde{x}_{ij}, \widetilde{x}_{io}\right)$.

Therefore, the relation (4.9) can be rewritten as $\mathbb{P}\left(\widetilde{h}_i \le -s_i^-\right) = 1-\alpha$. Now, according to the central limit theorem, we have $\mathbb{P}\left(\frac{\widetilde{h}_i - h_i}{\sigma_i^I(\lambda,\theta)} \le \frac{-s_i^- - h_i}{\sigma_i^I(\lambda,\theta)}\right) = 1-\alpha$. On the other hand, if we consider $\widetilde{z}_i = \frac{\widetilde{h}_i - h_i}{\sigma_i^I(\lambda,\theta)}$, then $\widetilde{z}_i \sim N(0,1)$. Hence, $\mathbb{P}\left(\widetilde{z}_i \le \frac{-s_i^- - h_i}{\sigma_i^I(\lambda,\theta)}\right) =$

$1-\alpha \Rightarrow \mathbb{P}\left(\widetilde{z}_i \leq \frac{s_i^-+h_i}{\sigma_i^I(\lambda,\theta)}\right)=\alpha, \Rightarrow \Phi\left(\frac{s_i^-+h_i}{\sigma_i^I(\lambda,\theta)}\right)=\alpha$, where Φ is a joint standard normal distribution function. Since Φ is an invertible function, then $\frac{s_i^-+h_i}{\sigma_i^I(\lambda,\theta)}=\Phi^{-1}(\alpha) \Rightarrow h_i+s_i^- - \Phi^{-1}(\alpha)\sigma_i^I(\lambda,\theta)=0$.

Eventually, the chance constraint of the ith input in model (4.8) is $\sum_{j=1}^{n}\lambda_j x_{ij} + s_i^- - \Phi^{-1}(\alpha)\sigma_i^I(\lambda,\theta)=\theta x_{io}$, where $v_i=\sum_{j=1}^{n}\sum_{k=1}^{n}\lambda_j\lambda_k cov(\widetilde{x}_{ij},\widetilde{x}_{ik})+\theta^2 var(\widetilde{x}_{io}) - 2\theta\sum_{j=1}^{n}\lambda_j cov(\widetilde{x}_{ij},\widetilde{x}_{io})$. Similarly, the chance constraint of the r^{th}output in model (4.8) is $\sum_{j=1}^{n}\lambda_j y_{rj} - s_r^+ + \Phi^{-1}(\alpha)\sigma_r^o(\lambda)=y_{ro}$, where $\sigma_r^o(\lambda)=\sum_{j=1}^{n}\times\sum_{k=1}^{n}\lambda_j\lambda_k cov(\widetilde{y}_{rj},\widetilde{y}_{rk})+var(\widetilde{y}_{ro})-2\sum_{j=1}^{n}\lambda_j cov(\widetilde{y}_{rj},\widetilde{y}_{ro})$ for all r.

Consequently, the deterministic form of the input-oriented stochastic CCR model (4.8) is set up as

$$\begin{aligned}
&\text{Min}\theta \qquad (4.10)\\
\text{s.t.}\quad &\sum_{j=1}^{n}\lambda_j x_{ij} + s_i^- - \Phi^{-1}(\alpha)\sigma_i^I(\lambda,\theta)=\theta x_{io}\ (\forall i),\\
&\sum_{j=1}^{n}\lambda_j y_{rj} - s_r^+ + \Phi^{-1}(\alpha)\sigma_r^o(\lambda)=y_{ro}\ (\forall r),\\
&s_i^- \geq 0,\quad s_r^+ \geq 0, i=1,\ldots,m\ (\forall r),\\
&\lambda_j \geq 0\ (\forall j).
\end{aligned}$$

Now, by defining the positive variables $v_i=\sigma_i^I(\lambda,\theta)$ and $u_r=\sigma_r^o(\lambda)$, model (4.10) can be converted into the following quadratic form:

$$\begin{aligned}
&\text{Min}\theta \qquad (4.11)\\
\text{s.t.}\quad &\sum_{j=1}^{n}\lambda_j x_{ij} + s_i^- - \Phi^{-1}(\alpha)v_i=\theta x_{io}\quad (\forall i),\\
&\sum_{j=1}^{n}\lambda_j y_{rj} - s_r^+ + \Phi^{-1}(\alpha)u_r=y_{ro}\quad (\forall r),\\
v_i &= \sum_{j=1}^{n}\sum_{k=1}^{n}\lambda_j\lambda_k cov(\widetilde{x}_{ij},\widetilde{x}_{ik})+\theta^2 var(\widetilde{x}_{io}) - 2\theta\sum_{j=1}^{n}\lambda_j cov(\widetilde{x}_{ij},\widetilde{x}_{io})\quad (\forall i),\\
u_r &= \sum_{j=1}^{n}\sum_{k=1}^{n}\lambda_j\lambda_k cov(\widetilde{y}_{rj},\widetilde{y}_{rk})+var(\widetilde{y}_{ro}) - 2\sum_{j=1}^{n}\lambda_j cov(\widetilde{y}_{rj},\widetilde{y}_{ro})\quad (\forall r).\\
&s_i^- \geq 0\ (\forall i),\quad s_r^+ \geq 0\ (\forall r), v_i \geq 0\ (\forall i), u_r \geq 0\ (\forall r),\\
&\lambda_j \geq 0\ (\forall j).
\end{aligned}$$

Definition 4.1. Firm o is stochastic efficient if and only if $\theta^* = 1$.

The following theorems are useful.

Theorem 4.2 Model (4.11) is feasible for all α.

Theorem 4.3 For each $\alpha \le 0.5$, the optimal solution of model (4.11) lies between 0 and 1, *i.e.* $0 \le \theta^* \le 1$.

In an analogous manner, the output-oriented stochastic CCR model can be set up as

$$\text{Max}\varphi \tag{4.12}$$

$$\text{s.t.} \quad \sum\nolimits_{j=1}^{n} \lambda_j x_{ij} + s_i^- - \Phi^{-1}(\alpha) v_i = x_{io} \ (\forall i),$$

$$\sum\nolimits_{j=1}^{n} \lambda_j y_{rj} - s_r^+ + \Phi^{-1}(\alpha) u_r = \varphi y_{ro} \ (\forall r),$$

$$v_i = \sum_{j=1}^{n} \sum_{k=1}^{n} \lambda_j \lambda_k cov(\widetilde{x}_{ij}, \widetilde{x}_{ik}) + var(\widetilde{x}_{io}) - 2 \sum_{j=1}^{n} \lambda_j cov(\widetilde{x}_{ij}, \widetilde{x}_{io}) \ (\forall i),$$

$$u_r = \sum_{j=1}^{n} \sum_{k=1}^{n} \lambda_j \lambda_k cov(\widetilde{y}_{rj}, \widetilde{y}_{rk}) + \varphi^2 var(\widetilde{y}_{ro}) - 2\varphi \sum_{j=1}^{n} \lambda_j cov(\widetilde{y}_{rj}, \widetilde{y}_{ro}) \ (\forall r),$$

$$s_i^- \ge 0 \ (\forall i), \quad s_r^+ \ge 0 \ (\forall r), v_i \ge 0 \ (\forall i), u_r \ge 0 \ (\forall r),$$

$$\lambda_j \ge 0 \ (\forall j).$$

Theorem 4.4 For each $\alpha \le 0.5$, the optimal solution of model (4.12) is $\varphi^* \ge 1$.

Theorem 4.5 For each $\alpha \le 0.5$, for the optimal solutions of models (4.11) and (4.12), we have $\theta^* = \frac{1}{\varphi^*}$.

4.3.2 *Error Structure*

In this section, we define the error structure of random variables that is used by Cooper et al. (1998) to transform a chance-constrained problem into a linear form. Sharpe (1963) and Kahane (1977) used a single-factor assumption of random variables in economics and finance. Assume that the inputs and outputs of each firm j ($j = 1, \ldots, n$) follow the following single-factor symmetric distribution:

$$\widetilde{x}_{ij} = x_{ij} + a_{ij} \, \widetilde{\xi}_{ij}, \quad \text{and} \quad \widetilde{y}_{rj} = y_{rj} + b_{rj} \widetilde{\epsilon}_{rj}, \tag{4.13}$$

where a_{ij}, b_{rj}, x_{ij}, and y_{rj} are non-negative and $\widetilde{\xi}_{ij}$ and $\widetilde{\epsilon}_{rj}$ are independent random variables with each one following standard normal distribution and are having symmetric input–output error structure. In addition, x_{ij} and y_{rj} are assumed to be the expected values, and a_{ij} and b_{rj} are the standard deviations of $\widetilde{x}_{ij}$ and $\widetilde{y}_{rj}$, respectively. Moreover, suppose that all the inputs and outputs are uncorrelated, *i.e.*, for each $j \neq k$:

$$cov(\widetilde{x}_{ij}, \widetilde{x}_{ik}) = 0 \ (\forall i), \text{and } cov(\widetilde{y}_{rj}, \widetilde{y}_{rk}) = 0 \ (\forall r). \tag{4.14}$$

From (4.13), we get $\widetilde{x}_{ij} \sim N\left(x_{ij}, \sigma^2 a_{ij}^2\right)$ and $\widetilde{y}_{rj} \sim N\left(y_{rj}, \sigma^2 b_{rj}^2\right)$. For each $j = 1, \ldots, n$, let $\widetilde{\xi}_i = \widetilde{\xi}_{ij}$ and $\widetilde{\epsilon}_r = \widetilde{\epsilon}_{rj}$.

Now, consider the ith input chance constraint of the model (4.8):

$$\mathbb{P}\left\{\sum\nolimits_{j=1}^{n} \lambda_j \widetilde{x}_{ij} \leq \theta \widetilde{x}_{io}\right\} \geq 1 - \alpha. \tag{4.15}$$

By replacing $\widetilde{h}_i = \sum\nolimits_{j=1}^{n} \lambda_j \widetilde{x}_{ij} - \theta \widetilde{x}_{io}$, and according to the normal distribution, the error structure, and the assumption of the uncorrelated inputs and outputs, we have $\widetilde{h}_i = \left(\sum\nolimits_{j=1}^{n} \lambda_j x_{ij} - \theta x_{io}\right) + \widetilde{\xi}_i\left(\sum\nolimits_{j=1}^{n} \lambda_j a_{ij} - \theta a_{io}\right)$, where $\widetilde{h}_i \sim N\left(\left(\sum\nolimits_{j=1}^{n} \lambda_j x_{ij} - \theta x_{io}\right), \sigma^2\left(\sum\nolimits_{j=1}^{n} \lambda_j a_{ij} - \theta a_{io}\right)^2\right)$. Hence, the deterministic form of constraint (4.15) becomes $\sum\nolimits_{j=1}^{n} \lambda_j x_{ij} - \Phi^{-1}(\alpha)\sigma\left|\sum\nolimits_{j=1}^{n} \lambda_j a_{ij} - \theta a_{io}\right| \leq \theta x_{io}$.

Similarly, the deterministic form of the rth output chance constraint of model (4.8) can be expressed as $\sum\nolimits_{j=1}^{n} \lambda_j y_{rj} + \Phi^{-1}(\alpha)\sigma\left|\sum\nolimits_{j=1}^{n} \lambda_j b_{rj} - \theta b_{ro}\right| \geq y_{ro}$.

Therefore, the deterministic form of model (4.8) becomes

$$\begin{aligned} & \text{Min}\theta \\ \text{s.t.} \quad & \sum\nolimits_{j=1}^{n} \lambda_j x_{ij} - \Phi^{-1}(\alpha)\sigma\left|\sum\nolimits_{j=1}^{n} \lambda_j a_{ij} - \theta a_{io}\right| \leq \theta x_{io} \ (\forall i), \\ & \sum\nolimits_{j=1}^{n} \lambda_j y_{rj} + \Phi^{-1}(\alpha)\sigma\left|\sum\nolimits_{j=1}^{n} \lambda_j b_{rj} - \theta b_{ro}\right| \geq y_{ro} \ (\forall r), \\ & \lambda_j \geq 0 \ (\forall j). \end{aligned} \tag{4.16}$$

Due to the existence of the absolute expression, model (4.16) is a non-linear model. To linearize this model, we consider the following transformations:

$$\begin{aligned} & \left|\sum\nolimits_{j=1}^{n} \lambda_j a_{ij} - \theta a_{io}\right| = p_i^+ + p_i^- \ (\forall i), \\ & \sum\nolimits_{j=1}^{n} \lambda_j a_{ij} - \theta a_{io} = p_i^+ - p_i^- \ (\forall i), \\ & p_i^+ p_i^- = 0 \ (\forall i), \end{aligned}$$

$$\left|\sum_{j=1}^{n}\lambda_j b_{rj} - \theta b_{ro}\right| = q_r^+ + q_r^- \ (\forall r),$$

$$\sum_{j=1}^{n}\lambda_j b_{rj} - \theta b_{ro} = q_r^+ - q_r^- \ (\forall r),$$

$$q_r^+ q_r^- = 0 \ (\forall r),$$

$$q_r^+ \geq 0,\ q_r^- \geq 0,\ p_i^+ \geq 0,\ p_i^- \geq 0 \ (\forall r, i).$$

The existence of constraints $p_i^+ p_i^- = 0$ and $q_r^+ q_r^- = 0$ implies that the problem (4.16) is a non-linear programming problem. If we ignore these constraints, the problem becomes linear. Moreover, if the linear programming problem has an optimal solution, then it has an optimal extreme point at which at least one of the values of $p_i^+ \ or \ p_i^-$ and $q_r^+ \ or \ q_r^-$ are zero. Consequently, if we use the simplex algorithm to solve this linear programming problem, we can find this optimal extreme point and we can, hence, avoid the constraints $p_i^+ p_i^- = 0$ and $q_r^+ q_r^- = 0$. Therefore, model (4.16) can be transformed into the following form:

$$\text{Min}\theta \tag{4.17}$$

$$\text{s.t.} \quad \sum_{j=1}^{n}\lambda_j x_{ij} - \Phi^{-1}(\alpha)\sigma\left(p_i^+ + p_i^-\right) \leq \theta x_{io} \ (\forall i),$$

$$\sum_{j=1}^{n}\lambda_j a_{ij} - \theta a_{io} = p_i^+ - p_i^- \ (\forall i),$$

$$\sum_{j=1}^{n}\lambda_j y_{rj} + \Phi^{-1}(\alpha)\sigma\left(q_r^+ + q_r^-\right) \geq y_{ro} \ (\forall r),$$

$$\sum_{j=1}^{n}\lambda_j b_{rj} - \theta b_{ro} = q_r^+ - q_r^- \ (\forall r),$$

$$\lambda_j \geq 0 \ (\forall j).$$

Theorem 4.6 Model (4.17) is feasible under all error levels of α.

Theorem 4.7 For any predetermined level of $\alpha \leq 0.5$, the stochastic efficiency score calculated from model (4.17) lies between 0 and 1, *i.e.* $0 \leq \theta^* \leq 1$.

Theorem 4.8 For any predetermined levels of α and α' with $\alpha \leq \alpha'$, we have $\theta^*(\alpha') \leq \theta^*(\alpha)$.

Note: It follows from Theorem (4.8) that if firm o is efficient at error level α', then it is efficient at any error level α, where $\alpha < \alpha'$. So, if a firm is efficient at the error level $\alpha' = 0.5$, then it is efficient for each error level $\alpha < 0.5$. It should be noted that model (4.17) at error level 0.5 becomes the deterministic CCR model with the mean data. Moreover, it follows from Theorem (4.8) that if firm o is inefficient at error level α', then for $\alpha < \alpha'$ it is also inefficient. So, if one unit is inefficient at the error level $\alpha = 0$, then it will be inefficient at any error level. Since $\Phi^{-1}(0) = -\infty$ but $\Phi^{-1}(0.001) \approx -3$, by evaluating each unit at the level $\alpha = 0.001$, one can always find units that are always inefficient.

4.3.3 Chance-Constrained DEA Model (P-Model)

So far, we have introduced the chance-constrained "E-model" in DEA. We now introduce another type of chance-constrained DEA model that is called the "P-model." In addition, the satisficing models based on the "P-model," which were introduced by Cooper et al. (1996), will be presented.

We now turn to show how the CCR model of Charnes et al. (1978) can be adapted in a chance-constrained programming context. Consider the following model:

$$\begin{aligned} \text{Max } & \mathbb{P}\left(\frac{\sum_{r=1}^{s} u_r \widetilde{y}_{ro}}{\sum_{i=1}^{m} v_i \widetilde{x}_{io}} \geq 1\right) \\ \text{s.t.} \quad & \mathbb{P}\left(\frac{\sum_{r=1}^{s} u_r \widetilde{y}_{rj}}{\sum_{i=1}^{m} v_i \widetilde{x}_{ij}} \leq 1\right) \geq 1 - \alpha_j \ (\forall j), \\ & u_r \geq 0, v_i \geq 0 \ (\forall r, i), \end{aligned} \tag{4.18}$$

where u_r and v_i are the virtual multipliers (or weights) to be determined by solving the above problem.

The constraints in model (4.18) are satisfied when $u_r = 0$ and $v_i > 0$, for all r and i. Hence, for continuous distributions, we can write:

$$\mathbb{P}\left(\frac{\sum_{r=1}^{s} u_r^* \widetilde{y}_{ro}}{\sum_{i=1}^{m} v_i^* \widetilde{x}_{io}} \leq 1\right) + \mathbb{P}\left(\frac{\sum_{r=1}^{s} u_r^* \widetilde{y}_{ro}}{\sum_{i=1}^{m} v_i^* \widetilde{x}_{io}} \geq 1\right) = 1$$

or

$$\mathbb{P}\left(\frac{\sum_{r=1}^{s} u_r^* \widetilde{y}_{ro}}{\sum_{i=1}^{m} v_i^* \widetilde{x}_{io}} \leq 1\right) = 1 - \alpha^* \geq 1 - \alpha_o,$$

where α^* is an optimal solution of model (4.18). In other words, α^* is the probability of achieving a value of at least unity with this choice of weights, and $1 - \alpha^*$ is, therefore, the probability of failing to achieve this value.

It should be noted that $\alpha_o \geq \alpha^*$ always holds since $1 - \alpha_o$ is prescribed in the constraint for $j = 0$ as the chance allowed for characterizing the $\widetilde{y}_{ro}$ and $\widetilde{x}_{io}$ values as inefficient. More formally, we introduce the following stochastic definition of efficiency:

Definition 4.2. Firm o is "stochastic efficient" if and only if $\alpha^* = \alpha_o$.

As we know, random variables are assumed to be managed by a known multivariate distribution. Therefore, we can consider the value of α^* even before data generation. If this value is too small, then the administrator can be notified. This means that there is a need to review firm o's improvement, which means that given the current conditions, the future performance will not be efficient.

4.3.4 Satisficing[1] Model

The following model is a generalization of the model (4.18):

$$\text{Max } \mathbb{P}\left(\frac{\sum_{r=1}^{s} u_r \widetilde{y}_{ro}}{\sum_{i=1}^{m} v_i \widetilde{x}_{io}} \geq \beta_o\right) \tag{4.19}$$

$$\text{s.t.} \quad \mathbb{P}\left(\frac{\sum_{r=1}^{s} u_r \widetilde{y}_{rj}}{\sum_{i=1}^{m} v_i \widetilde{x}_{ij}} \leq \beta_o\right) \geq 1 - \alpha_j \; (\forall j),$$

$$u_r \geq 0, v_i \geq 0 \;\; (\forall r, i),$$

where β_o represents the "aspiration level," either imposed by an outside authority, as in the budget control model by Stedry (1960), or by the administrator of some activities, as in Simon's (1957) "satisficing" approach. Simon (1957) is the first to present the theory of bounded rationality. He described decision-making as a search process guided by aspiration levels. An aspiration level is a value of a goal variable, which must be reached or surpassed by a satisfactory decision alternative. Simon (1957) used the concept of "satisficing" to explain the behavior of decision-makers in various situations when an optimal solution cannot be determined.

The above models are very general and are mainly intended for conceptual clarity. For the calculations and applicability of these models, a deterministic equivalent form is required. To this end, we assume that all inputs and outputs are random variables with each following a multivariate normal distribution with known parameters. The choice of a normal multivariate distribution may be a bit restrictive, but using transformations, other types of distributions can be transformed into normal distribution approximately.

Subsequently, using Cooper et al.'s (1996) suggested procedure, we transform model (4.19) into a deterministic model. From the constraint (4.19), we have

$$\mathbb{P}\left(U\widetilde{y}_j - \beta_j V\widetilde{x}_j \leq 0\right) \geq 1 - \alpha_j$$

$$\Rightarrow \mathbb{P}\left(\frac{U\widetilde{y}_j - \beta_j V\widetilde{x}_j - \left(Uy_j - \beta_j Vx_j\right)}{\sigma_{U\widetilde{y}_j - \beta_j V\widetilde{x}_j}} \leq \frac{-\left(Uy_j - \beta_j Vx_j\right)}{\sigma_{U\widetilde{y}_j - \beta_j V\widetilde{x}_j}}\right)$$
$$\geq 1 - \alpha_j, \tag{4.20}$$

where $x_j = \mathbb{E}\left[\widetilde{x}_j\right], y_j = \mathbb{E}\left[\widetilde{y}_j\right]$, and $\sigma_{U\widetilde{y}_j - \beta_j V\widetilde{x}_j} = \sqrt{Var\left(U\widetilde{y}_j - \beta_j V\widetilde{x}_j\right)}$.

Now, we introduce a new random variable $\widetilde{\xi}_j$, which we present as

[1] The word "satisficing" is a combination of the two words "satisfying" and "sufficient." Thus, the concept of the new term is "satisfactory enough."

$$\widetilde{\xi}_j = \frac{U\widetilde{y}_j - \beta_j V\widetilde{x}_j - (Uy_j - \beta_j Vx_j)}{\sigma_{U\widetilde{y}_j - \beta_j V\widetilde{x}_j}} \ (\forall j). \tag{4.21}$$

Here $\widetilde{\xi}_j$ follows the standard normal probability distribution and its direct substitution in (4.20) yields

$$\mathbb{P}\left(\widetilde{\xi}_j \leq \frac{-(Uy_j - \beta_j Vx_j)}{\sigma_{U\widetilde{y}_j - \beta_j V\widetilde{x}_j}}\right) \geq 1 - \alpha_j \ (\forall j). \tag{4.22}$$

According to the standard normal distribution function, (4.22) can be expressed as

$$\Phi\left(\frac{-(Uy_j - \beta_j Vx_j)}{\sigma_{U\widetilde{y}_j - \beta_j V\widetilde{x}_j}}\right) \geq 1 - \alpha_j \ (\forall j). \tag{4.23}$$

Since $\Phi(\cdot)$ is an invertible and strictly increasing function, then (4.23) can be written as

$$\frac{-(Uy_j - \beta_j Vx_j)}{\sigma_{U\widetilde{y}_j - \beta_j V\widetilde{x}_j}} \geq \Phi^{-1}(1 - \alpha_j) \ (\forall j). \tag{4.24}$$

In the spirit of Charnes and Cooper (1963), we introduce non-negative "spacer variables" η_j and replace (4.23) with the following two inequalities:

$$\begin{aligned} &(Uy_j - \beta_j Vx_j) - \Phi^{-1}(\alpha_j)\eta_j \leq 0 \ (\forall j), \\ &C_{\alpha_j}\left[U^t\Sigma_{yy}U + \beta_j^2 V^t\Sigma_{xx}V - 2\beta_j U^t\Sigma_{yx}V - \eta_j^2\right] \leq 0 \ (\forall j), \end{aligned} \tag{4.25}$$

where

$$C_{\alpha_j} = \begin{cases} 1 & \text{if } \alpha_j < 0.5; \\ 0 & \text{if } \alpha_j = 0.5; \\ -1 & \text{if } \alpha_j > 0.5. \end{cases}$$

Using the deterministic constraints (4.25), model (4.19) can alternatively be set up as

$$\text{Max } \mathbb{P}\left(\frac{U\widetilde{y}_o}{V\widetilde{x}_o} \geq \beta_o\right) \tag{4.26}$$

$$\text{s.t.} \quad (Uy_j - \beta_j Vx_j) - \Phi^{-1}(\alpha_j)\eta_j \leq 0 \ (\forall j),$$

$$C_{\alpha_j}\left[U^t\Sigma_{yy}U + \beta_j^2 V^t\Sigma_{xx}V - 2\beta_j U^t\Sigma_{yx}V - \eta_j^2\right] \leq 0 \ (\forall j),$$

$$U_r, V_i, \eta_j \geq 0 \ (\forall r, i, j).$$

The constraints in (4.26) are deterministic but due to the existence of the random variables $\widetilde{x}_o$ and $\widetilde{y}_o$ in the objective function, this problem is not yet deterministic. Consider below a model that is equivalent to (4.26):

$$\text{Max} \quad \gamma \tag{4.27}$$

$$\text{s.t.} \quad \mathbb{P}\left(\frac{U\widetilde{y}_o}{V\widetilde{x}_o} \geq \beta_o\right) \geq \gamma,$$

$$(Uy_j - \beta_j Vx_j) - \Phi^{-1}(\alpha_j)\eta_j \leq 0 \ (\forall j),$$

$$C_{\alpha_j}\left[U^t\Sigma_{yy}U + \beta_j^2 V^t\Sigma_{xx}V - 2\beta_j U^t\Sigma_{yx}V - \eta_j^2\right] \leq 0 \ (\forall j),$$

$$U_r, V_i, \eta_j \geq 0 \ (\forall r, i, j).$$

Similarly, as with the previous constraints, the first constraint in (4.26) can be transformed into the following deterministic form:

$$\mathbb{P}\left(\frac{U\widetilde{y}_o}{V\widetilde{x}_o} \geq \beta_o\right) \geq \gamma \Rightarrow \mathbb{P}(\ U\widetilde{y}_o \geq \beta_o V\widetilde{x}_o\ ) \geq \gamma$$

$$\Rightarrow \mathbb{P}(\ \beta_o V\widetilde{x}_o - U\widetilde{y}_o \leq 0\ ) \geq \gamma$$

$$\Rightarrow \mathbb{P}\left(\frac{\beta_o V\widetilde{x}_o - U\widetilde{y}_o - (\beta_o Vx_o - Uy_o)}{\sigma_{\beta_o V\widetilde{x}_o - U\widetilde{y}_o}} \leq \frac{-(\beta_o Vx_o - Uy_o)}{\sigma_{\beta_o V\widetilde{x}_o - U\widetilde{y}_o}}\right) \geq \gamma$$

$$\Rightarrow \Phi\left(\frac{Uy_o - \beta_o Vx_o}{\sigma_{\beta_o V\widetilde{x}_o - U\widetilde{y}_o}}\right) \geq \gamma$$

$$\Rightarrow \quad \frac{Uy_o - \beta_o Vx_o}{\sigma_{\beta_o V\widetilde{x}_o - U\widetilde{y}_o}} \geq \Phi^{-1}(\gamma).$$

Model (4.28), which is equivalent to (4.27), will now be transformed as

$$\text{Max} \quad \gamma \tag{4.28}$$
$$\text{s.t.} \quad \frac{Uy_o - \beta_o Vx_o}{\sigma_{\beta_o V\tilde{x}_o - U\tilde{y}_o}} \geq \Phi^{-1}(\gamma),$$
$$(Uy_j - \beta_j Vx_j) - \Phi^{-1}(\alpha_j)\eta_j \leq 0 \ (\forall j),$$
$$C_{\alpha_j}\left[U^t\Sigma_{yy}U + \beta_j^2 V^t\Sigma_{xx}V - 2\beta_j U^t\Sigma_{yx}V - \eta_j^2\right] \leq 0 \ (\forall j),$$
$$U_r, V_i, \eta_j \geq 0 \ (\forall r, i, j).$$

Although problem (4.28) is deterministic, it is not easy to work with because the existence of the denominator in the first constraint makes this a non-convex programming problem (Cooper et al., 1996). To solve this problem, we introduce a variable θ as $\theta = \Phi^{-1}(\gamma) \Rightarrow \gamma = \Phi(\theta)$. By replacing θ in the model (4.28), we obtain the following model:

$$\text{Max} \quad \Phi(\theta) \tag{4.29}$$
$$\text{s.t.} \quad \frac{Uy_o - \beta_o Vx_o}{\sigma_{\beta_o V\tilde{x}_o - U\tilde{y}_o}} \geq \theta,$$
$$(Uy_j - \beta_j Vx_j) - \Phi^{-1}(\alpha_j)\eta_j \leq 0 \ (\forall j),$$
$$C_{\alpha_j}\left[U^t\Sigma_{yy}U + \beta_j^2 V^t\Sigma_{xx}V - 2\beta_j U^t\Sigma_{yx}V - \eta_j^2\right] \leq 0 \ (\forall j),$$
$$U_r, V_i, \eta_j \geq 0 \ (\forall r, i, j).$$

Since $\Phi(\Theta)$ is a strictly increasing function of θ, we can maximize θ instead of $\Phi(\theta)$. So, we can solve the following model:

$$\text{Max} \quad \theta \tag{4.30}$$
$$\text{s.t.} \quad \frac{Uy_o - \beta_o Vx_o}{\sigma_{\beta_o V\tilde{x}_o - U\tilde{y}_o}} \geq \theta,$$
$$(Uy_j - \beta_j Vx_j) - \Phi^{-1}(\alpha_j)\eta_j \leq 0 \ (\forall j),$$
$$C_{\alpha_j}\left[U^t\Sigma_{yy}U + \beta_j^2 V^t\Sigma_{xx}V - 2\beta_j U^t\Sigma_{yx}V - \eta_j^2\right] \leq 0 \ (\forall j),$$
$$U_r, V_i, \eta_j \geq 0 \ (\forall r, i, j).$$

It is now easy to see that model (4.30) is equivalent to the following problem:

$$\text{Max} \quad \frac{Uy_o - \beta_o Vx_o}{\sigma_{\beta_o V\widetilde{x_o} - U\widetilde{y_o}}} \tag{4.31}$$

$$\text{s.t.} \quad (Uy_j - \beta_j Vx_j) - \Phi^{-1}(\alpha_j)\eta_j \leq 0 \ (\forall j),$$

$$C_{\alpha_j}\left[U^t\Sigma_{yy}U + \beta_j^2 V^t\Sigma_{xx}V - 2\beta_j U^t\Sigma_{yx}V - \eta_j^2\right] \leq 0 \ (\forall j),$$

$$U_r, V_i, \eta_j \geq 0 \ (\forall r, i, j)$$

since $(Uy_o - \beta_o Vx_j)/\sigma_{\beta_o V\widetilde{x_o} - U\widetilde{y_o}}$ is bounded by $\Phi^{-1}(\alpha_o)$, $\frac{Uy_o - \beta_o Vx_o}{\sigma_{\beta_o V\widetilde{x_o} - U\widetilde{y_o}}} \leq \Phi^{-1}(\alpha_o)$.

In the spirit of the Charnes and Cooper's (1963) approach, we introduce a positive variable ϖ, and because $\frac{Uy_o - \beta_o Vx_o}{\sigma_{\beta_o V\widetilde{x_o} - U\widetilde{y_o}}} \leq \frac{(Uy_o - \beta_o Vx_o)}{\varpi} \leq \Phi^{-1}(\alpha_o)$, we deduce that $\sigma_{\beta_o V\widetilde{x_o} - U\widetilde{y_o}} \geq \varpi \Rightarrow \sigma^2_{\beta_o V\widetilde{x_o} - U\widetilde{y_o}} \geq \varpi^2$.

Hence, model (4.31) is equivalent to the following problem:

$$\text{Max} \quad \frac{(Uy_o - \beta_o Vx_o)}{\varpi} \tag{4.32}$$

$$\text{s.t.} \quad \beta_o^2 \, V^t\Sigma_{x_o x_o}V + U^t\Sigma_{y_o y_o}U - 2\beta_o U^t\Sigma_{y_o x_o}V - \varpi^2 \geq 0,$$

$$(Uy_j - \beta_j Vx_j) - \Phi^{-1}(\alpha_j)\eta_j \leq 0 \ (\forall j),$$

$$C_{\alpha_j}\left[U^t\Sigma_{yy}U + \beta_j^2 V^t\Sigma_{xx}V - 2\beta_j U^t\Sigma_{yx}V - \eta_j^2\right] \leq 0 \ (\forall j),$$

$$U_r, V_i, \eta_j \geq 0 \ (\forall r, i, j).$$

The objective function of model (4.32) is in fractional form. Using Charnes and Cooper's (1962) transformation approach, we consider the following transformations $t{:=}\frac{1}{\varpi}, \zeta_j{:=}t\eta_j, \ \mu{:=}tU, \ \upsilon{:=}tV$, to convert the model (4.32) into an equivalent quadratic programming problem:

$$\text{Max} \quad (\mu y_o - \beta_o \upsilon x_o) \tag{4.33}$$

$$\text{s.t.} \quad \mu^t\Sigma_{y_o y_o}\mu + \beta_o^2 \upsilon^t\Sigma_{x_o x_o}\upsilon - 2\beta_o \mu^t\Sigma_{y_o x_o}\upsilon \geq 1$$

$$(\mu^t y_j - \beta_j \upsilon^t x_j) - \Phi^{-1}(\alpha_j)\zeta_j \leq 0 \ (\forall j),$$

$$C_{\alpha_j}\left[\mu^t\Sigma_{yy}\mu + \beta_j^2 \upsilon^t\Sigma_{xx}\upsilon - 2\beta_j \mu^t\Sigma_{yx}\upsilon - \zeta_j^2\right] \leq 0 \ (\forall j),$$

$$\lambda, \upsilon, \zeta \geq 0 \ (\forall r, i, j).$$

We conclude this section with a theorem that shows the relationship between the solutions of models (4.33) and (4.19).

Theorem 4.9 For $\beta_o = 1$ ($\beta_o < 1$), let (μ^*, v^*, ζ^*) and (W^*, U^*, V^*) be the optimal solutions of models (4.33) and (4.19), respectively. Then, $\Phi(\mu^* y_o - \beta_o v^* x_o) = \mathbb{P}\left(\frac{U^*\widetilde{y_o}}{V^*\widetilde{x_o}} \geq \beta_o\right)$. Furthermore, firm o is stochastically (satisficing) efficient if and only if $\Phi(\mu^* y_o - \beta_o \lambda^* x_o) = \alpha_0$.

Proof From models (4.29) and (4.28), we have $\Phi(\mu^* y_o - \beta_o \lambda^* z_o) = \gamma^*$, where γ^* is the optimal solution for problem (4.28). Since problem (4.28) is equivalent to the problem (4.19), then $\gamma^* = \mathbb{P}\left(\frac{U^*\widetilde{y_o}}{W^*\widetilde{z_o}} \geq \beta_o\right)$, and, consequently, $\Phi(\mu^* y_o - \beta_o \lambda^* z_o) = \mathbb{P}\left(\frac{U^*\widetilde{y_o}}{W^*\widetilde{z_o}} \geq \beta_o\right)$. ∎

4.3.5 Transformation to a Linear Programming Problem

Stochastic DEA requires a large number of parameters. The practical application of this technique can be greatly enhanced by considering some assumptions that help decrease the computational estimation of parameters and the transformation of the programming problem with chance constraints into its linear programming equivalent. One of these assumptions is described in this section.

Like Cooper et al. (1996), we assume that the input and output random variables follow a single-factor symmetric distribution. Sharpe (1963) and Kahane (1977) used this single-factor assumption in economics and finance. Suppose that:

$$\widetilde{x}_{ij} = x_{ij} + a_{ij}\,\xi,$$
$$\widetilde{y}_{rj} = y_{rj} + b_{rj}\,\epsilon,$$

where x_{ij}, y_{rj}, a_{ij}, and b_{rj} are positive, and ξ and ϵ are the random variables with each one following the standard normal distribution. In addition, x_{ij} and y_{rj} are assumed to be the expected values and a_{ij} and b_{rj} are the standard deviations of $\widetilde{x}_{ij}$ and $\widetilde{y}_{rj}$, respectively.

For the constraint in model (4.19), we have:

$$\mathbb{P}\left(\frac{U\widetilde{y}_j}{V\widetilde{x}_j} \leq \beta_j\right) \geq 1 - \alpha_j \Rightarrow \mathbb{P}\left(U\widetilde{y}_j - \beta_j V\widetilde{x}_j \leq 0\right) 1 - \alpha_j. \tag{4.34}$$

Now, by considering $\widetilde{h} = U\widetilde{y}_j - \beta_j V\widetilde{x}_j$ and the single-factor assumption, we obtain $\widetilde{h} = Uy_j - \beta_j Vx_j + Ub_j\epsilon - \beta_j Va_j\xi$, where $\widetilde{h} \sim N\left(Uy_j - \beta_j Vx_j, \left(Ub_j - \beta_j Va_j\right)^2\right)$. Hence, from (4.34), it can be seen that $\mathbb{P}\left(\frac{U\widetilde{y}_j - \beta_j V\widetilde{x}_j - \left(Uy_j - \beta_j Vx_j\right)}{\left|Ub_j - \beta_j Va_j\right|} \leq \frac{-\left(Uy_j - \beta_j Vx_j\right)}{\left|Ub_j - \beta_j Va_j\right|}\right) \geq 1 - \alpha_j$, in which

$\frac{U\tilde{y}_j-\beta_j V\tilde{x}_j-(Uy_j-\beta_j Vx_j)}{|Ub_j-\beta_j Va_j|} \sim N(0,1)$, and $\Phi\left(\frac{-(Uy_j-\beta_j Vx_j)}{|Ub_j-\beta_j Va_j|}\right) \geq 1-\alpha_j \Rightarrow$ $\frac{-(Uy_j-\beta_j Vx_j)}{|Ub_j-\beta_j Va_j|} \geq \Phi^{-1}(1-\alpha_j)$. Eventually, the deterministic form of the constraint in (4.19) is $\beta_j Vx_j - Uy_j \geq \Phi^{-1}(1-\alpha_j)|Ub_j - \beta_j Va_j|$.

In an analogous manner, model (4.19) can be transformed into the following form:

$$\begin{aligned} \text{Max} \quad & (wy_o - \beta_o vx_o) \\ \text{s.t.} \quad & |\beta_o va_o - wb_o| \geq 1, \\ & (\beta_j vx_j - wy_j) \geq \Phi^{-1}(1-\alpha_j)|\beta_j va_j - wb_j| \ (\forall j), \\ & w, v \geq 0. \end{aligned} \tag{4.35}$$

Since the constraints in the model (4.35) include the absolute function, it is not a linear programming problem. Therefore, we consider the following transformations:

$$\begin{aligned} & |\beta_j va_j - wb_j| = p_j^+ + p_j^-, \\ & \beta_j va_j - wb_j = p_j^+ - p_j^-, \\ & p_j^+ p_j^- = 0, \\ & p_j^+, p_j^- \geq 0, \end{aligned}$$

and replace them in (4.35) to obtain the following quadratic programming:

$$\begin{aligned} \text{Max} \quad & (wy_o - \beta_o vx_o) \\ \text{s.t.} \quad & p_o^+ + p_o^- \geq 1, \\ & (\beta_o vx_j - wy_j) \geq \Phi^{-1}(1-\alpha_j)\left(p_j^+ + p_j^-\right) \ (\forall j), \\ & \beta_j va_j - wb_j = p_j^+ - p_j^- \ (\forall j), \\ & p_j^+ p_j^- = 0 \ (\forall j), \\ & w, v, p^+, p^- \geq 0. \end{aligned} \tag{4.36}$$

Due to the existence of the constraint $p_j^+ p_j^- = 0$, problem (4.36) is a non-linear programming problem. If we ignore this constraint, the problem becomes linear. Moreover, if the linear programming problem has an optimal solution, then it has an optimal extreme point at which at least one of the values of p_j^+ or p_j^- is zero. Consequently, if we use the simplex algorithm to solve this linear programming problem, we can find this optimal extreme point and can hence avoid the constraint $p_j^+ p_j^- = 0$. So, we solve the following linear programming problem:

$$\begin{aligned} \text{Max} \quad & (wy_o - \beta_o vx_o) \\ \text{s.t.} \quad & p_o^+ + p_o^- \geq 1, \\ & (\beta_j vx_j - wy_j) \geq \Phi^{-1}(1-\alpha_j)\left(p_j^+ + p_j^-\right) \ (\forall j), \\ & \beta_j va_j - wb_j = p_j^+ - p_j^- \ (\forall j), \\ & w, v, p^+, p^- \geq 0. \end{aligned} \tag{4.37}$$

References

Banker, R. D. (1986). *Stochastic data envelopment analysis*. Carnegie-Mellon University.

Banker, R. (1989). Econometric estimation and data envelopment analysis. *Research in Governmental and Nonprofit Accounting, 5*, 231–343.

Banker, R. D. (1993). Maximum likelihood, consistency and data envelopment analysis: A statistical foundation. *Management Science, 39*(10), 1265–1273.

Banker, R. D. (1996). Hypothesis tests using data envelopment analysis. *Journal of Productivity Analysis, 7*(2/3), 139–160.

Banker, R. D., & Chang, H. (1995). A Monte Carlo estimation of hypothesis tests for differences in efficiencies. *International Journal of Production Economics, 39*, 37–54.

Banker, R. D., Chang, H., & Sinha, K. K. (1994). *Statistical tests to evaluate the separability or substitutability of inputs to a production system*. University of Minnesota. Working Paper.

Banker, R. D., Kotarac, K., & Neralic, L. (2015). Sensitivity and stability in stochastic data envelopment analysis. *Journal of Operational Research Society, 66*, 134–147.

Banker, R. D., & Maindiratta, A. (1992). Maximum likelihood estimation of monotone and concave production frontiers. *Journal of Productivity Analysis, 3*, 401–415.

Charles, V., & Cornillier, F. (2017). Value of the stochastic efficiency in data envelopment analysis. *Expert Systems with Applications, 81*, 349–357.

Charles, V., Tsolas, I. E., & Gherman, T. (2018). Satisficing data envelopment analysis: A Bayesian approach for peer mining in the banking sector. *Annals of Operations Research, 269*(1–2), 81–102.

Charles, V., & Udhayakumar, A. (2012). Genetic algorithm for chance constrained reliability stochastic optimisation problems. *International Journal of Operational Research, 14*(4), 417–432.

Charnes, A., & Cooper, W. W. (1962). Programming with linear fractional functionals. *Naval Research Logistics (NRL), 9*(3–4), 181–186.

Charnes, A., & Cooper, W. W. (1963). Deterministic equivalents for optimizing and satisficing under chance constraints. *Operations Research, 11*(1), 18–39.

Charnes, A., Cooper, W. W., & Rhodes, E. (1978). Measuring the efficiency of decision making units. *European Journal of Operational Research, 2*(6), 429–444.

Cooper, W. W., Huang, Z., Lelas, V., Li, S. X., & Olesen, O. B. (1998). Chance constrained programming formulations for stochastic characterizations of efficiency and dominance in DEA. *Journal of Productivity Analysis, 9*(1), 53–79.

Cooper, W., Huang, Z., & Li, S. X. (1996). Chapter 13 Satisficing DEA models under chance constraints. *Annals of Operations Research, 66*(4), 279–295.

Grosskopf, S. (1996). Statistical inference and nonparametric efficiency: A selective survey. *Journal of Productivity Analysis, 7*(2–3), 161–176.

Horrace, W. C., & Schmidt, P. (1996). Confidence statements for efficiency estimates from stochastic frontier models. *Journal of Productivity Analysis, 7*(2), 257–282.

Kahane, Y. (1977). Determination of the product mix and the business policy of an insurance company—a portfolio approach. *Management Science, 23*(10), 1060–1069.

Land, K. C., Lovell, C. K., & Thore, S. (1993). Chance-constrained data envelopment analysis. *Managerial and Decision Economics, 14*(6), 541–554.

Olesen, O. B., & Petersen, N. C. (1995). Chance constrained efficiency evaluation. *Management Science, 41*(3), 442–457.

Olesen, O. B., & Petersen, N. C. (2016). Stochastic data envelopment analysis—A review. *European Journal of Operational Research, 251*(1), 2–21.

Sengupta, J. K. (1990). Transformations in stochastic DEA models. *Journal of Econometrics, 46* (1–2), 109–123.

Sengupta, J. K. (2000). Efficiency analysis by stochastic data envelopment analysis. *Applied Economics Letters, 7*(6), 379–383.

Sharpe, W. F. (1963). A simplified model for portfolio analysis. *Management Science, 9*(2), 277–293.

Simar, L. (1996). Aspects of statistical analysis in DEA-type frontier models. *Journal of Productivity Analysis, 7*(2), 177–185.

Simar, L., & Wilson, P. W. (1998). Sensitivity analysis of efficiency scores: How to bootstrap in nonparametric frontier models. *Management Science, 44*(1), 49–61.

Simon, H. A. (1957). *Models of man* (p. 20). Wiley.

Stedry, A. C. (1960). *Budget control and cost behavior*. Prentice-Hall.

Udhayakumar, A., Charles, V., & Kumar, M. (2011). Stochastic simulation based genetic algorithm for chance constrained data envelopment analysis problems. *Omega, 39*(4), 387–397.

Chapter 5
Stochastic Network Data Envelopment Analysis

Most real-life production processes are multi-stage in nature. Characterization of such processes via concepts such as technical efficiency is considered important to firm managers for the stage-specific analysis of their business decisions in improving their performance. Therefore, it is imperative to estimate the efficiency of a firm not only for the network production system but also for its sub-processes to locate the sources of inefficiency. In this chapter, we deal with production processes characterized by a two-stage network structure that links their stage-specific processes with intermediate products (measures). In this two-stage production process, the first stage uses input resources to produce intermediate products, which are all, in turn, used as inputs in the second stage to produce final outputs.

To evaluate the relative efficiency of two-stage production processes, there are two major approaches: cooperative and non-cooperative. Assume a two-stage process in which one stage is considered the leader and the other the follower. In non-cooperative approaches, we first optimize the leader, and then, the efficiency of the follower is maximized (Liang et al., 2008). In contrast, in cooperative or centralized approaches, the importance of both stages is the same and the performance of the overall production system is maximized. For example, Seiford and Zhu (1999) employed a standard DEA approach to evaluate the profitability and marketability of US commercial banks without considering the intermediate measures (see also Yu & Shi, 2014). Kao and Hwang (2008) considered each stage as independent and evaluated the efficiency of the whole system as the product of the stage efficiencies. Liang et al. (2006) introduced several two-stage network models according to the concepts of non-cooperative and cooperative games theories. Furthermore, the optimal system design (OSD) network DEA models were proposed by Wei and Chang (2011) to help the firms or DMUs to find the optimal portfolios of inputs and outputs in terms of profit maximization. Furthermore, the two-stage approach has also been used to evaluate scale elasticity of the production systems along with its stages to locate the sources of scale economies (Kao & Hwang, 2011; Sahoo et al., 2014a, b).

A. Amirteimoori et al., *Stochastic Benchmarking*, International Series in Operations Research & Management Science 317,
https://doi.org/10.1007/978-3-030-89869-4_5

In the above-mentioned network models, the variability and uncertainty in the data have been ignored. Although some researchers have considered uncertainty in terms of fuzzy inputs and outputs (see, *e.g.*, Azadi et al., 2015; Dotoli et al., 2015; Hahn et al., 2016; Kao & Liu, 2011; Khalili-Damghani & Shahmir, 2015; Liu, 2014; Lozano, 2014; Lu, 2015; among others), there is little work on two-stage network DEA models that consider stochastic data (see Simar & Wilson, 2007; Banker & Natarajan, 2008; Banker et al., 2019). Zhou et al. (2017) proposed stochastic network DEA models under the centralized control organization mechanism. Their model was based on chance-constrained programming wherein they utilized the concept of expected value optimization, and hence, it falls under the class of "E-Models" (Charnes & Cooper, 1963). Izadikhah and Saen (2018) also proposed an alternative stochastic two-stage DEA model in the presence of undesirable data.

In this chapter, the game theory method by Liang et al. (2006) is used to measure the relative efficiency of a two-member supply chain. The authors used the leader–follower (non-cooperative) and cooperative (centralized) games to develop DEA models. We focus on production units with a two-stage structure with random inputs, intermediate measures, and final outputs.

Our first aim is to incorporate satisficing DEA and non-cooperative game theory concepts into the network processes with a two-stage structure. To this end, we focus on the class of "P-Models" introduced by Charnes and Cooper (1963) and propose a stochastic two-stage DEA model, namely a satisficing two-stage DEA model, according to the concept of the non-cooperative game theory (Leader–Follower). As a matter of fact, by using this methodology, we can replace deterministic concepts such as "efficient" and "inefficient" with concepts such as "stochastic efficient" and "stochastic inefficient" or "satisficing efficient" and "satisficing inefficient." Our second aim is to incorporate the chance-constrained DEA and cooperative game theory concepts into the network processes with a two-stage structure. Toward this end, we utilize the centralized two-stage model proposed by Liang et al. (2008) and the stochastic DEA proposed by Huang and Li (1996) and Cooper et al. (1998) to develop a stochastic centralized two-stage DEA model.

5.1 Stochastic Two-Stage Network DEA Models

Figure 5.1 describes a two-stage production process, where each DMU is composed of two sub-DMUs in a series relationship, and the sub-DMUs in Stage 2 consume the intermediate products of the sub-DMUs in Stage 1 as inputs. Each $DMU_{j \in \{1,2,\ldots,n\}}$ in Stage 1 uses the random input vector $\widetilde{x}_j = \left(\widetilde{x}_{1j}, \widetilde{x}_{2j}, \ldots, \widetilde{x}_{Mj}\right)^T$ to produce the random intermediate product vector $\widetilde{z}_j = \left(\widetilde{z}_{1j}, \widetilde{z}_{2j}, \ldots, \widetilde{z}_{Dj}\right)^T$, which are then used in Stage 2 as inputs to produce the final random output vector $\widetilde{y}_j = \left(\widetilde{y}_{1j}, \widetilde{y}_{2j}, \ldots, \widetilde{y}_{Sj}\right)^T$.

Note that if we look at the system as a black box with $\widetilde{x}_j$ as input and $\widetilde{y}_j$ as output and ignore the intermediate products, the stochastic CCR model can be used to calculate the stochastic efficiency score of the whole process for DMU_o (Zhou et al.,

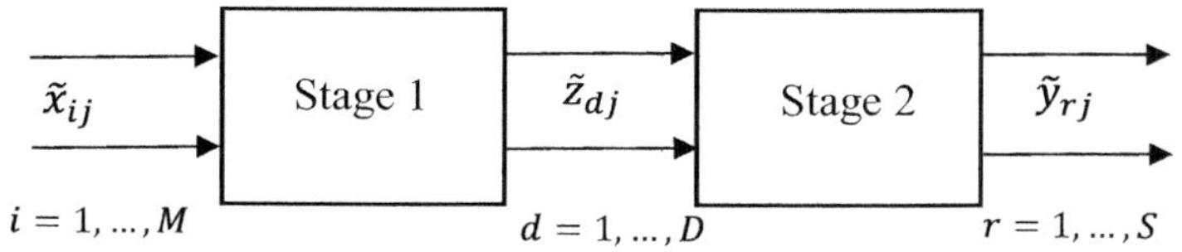

Fig. 5.1 Two-stage production process

2017). However, as Kao and Liu (2011) pointed out, this would be an invalid approach to measuring the performance of two-stage processes as it obscures the inefficiencies associated with various stages.

5.2 The Non-cooperative Model

In this section, we present the stochastic two-stage network DEA model according to the concepts of the non-cooperative game theory (Leader–Follower), which were utilized by Liang et al. (2006) to develop the models for performance evaluation.

5.2.1 *Stage 1 as the Leader and Stage 2 as the Follower*

Here, we assume that the first stage is the leader and the second stage is the follower. Hence, the importance of the first stage is higher than that of the second stage. Program (5.1) is used to evaluate the efficiency of the DMU$_o$ in Stage 1 (leader).

$$e_{1l}^{*(o)} = \text{Max } \mathbb{P}\left(\frac{W\tilde{z}_o}{V\tilde{x}_o} \geq \beta_o\right) \tag{5.1}$$

$$\text{s.t.} \quad \mathbb{P}\left(\frac{W\tilde{z}_j}{V\tilde{x}_j} \leq \beta_j\right) \geq 1 - \alpha_j \ (\forall j),$$

$$W, V \geq 0.$$

Cooper et al. (1996) obtained the deterministic form of the following stochastic programming program (see Sect. 4.3.4):

$$e_{1l}^{*(o)} = \text{Max } (\lambda z_0 - \beta_o \upsilon x_0) \tag{5.2}$$

$$\text{s.t.} \quad \lambda^t \Sigma_{z_0 z_o} \lambda + \beta_o^2 \upsilon^t \Sigma_{x_0 x_0} \upsilon - 2\beta_o \lambda^t \Sigma_{z_0 x_0} \upsilon \geq 1,$$

$$(\lambda z_j - \beta_j \upsilon x_j) - \Phi^{-1}(\alpha_j)\xi_j \leq 0 \ (\forall j),$$

$$C_{\alpha_j}\left[\lambda^t\Sigma_{zz}\lambda+\beta_j^2 v^t\Sigma_{xx}v-2\beta_j\lambda^t\Sigma_{zx}v-\xi_j^2\right]\le 0\ (\forall j),$$
$$\lambda, v, \xi \ge 0.$$

The relationship between the optimal solutions obtained from programs (5.1) and (5.2) is presented in Theorem 5.1.

Theorem 5.1 Suppose $\beta_{jo}=\beta_o=1$ ($\beta_{jo}=\beta_o<1$). Let (λ^*, v^*, ξ^*) and (W^*, V^*) be the optimal solutions obtained from programs (5.2) and (5.1), respectively. Then, $\Phi(\lambda^* z_0-\beta_o v^* x_0)=\mathbb{P}\left(\frac{W^*\widetilde{z}_o}{V^*\widetilde{x}_o}\ge\beta_o\right)$. Furthermore, DMU_o is stochastically (satisficing) efficient if and only if $\Phi(\lambda^* z_0-\beta_o v^* x_0)=\alpha_0$.

Proof The proof is similar to the proof of Theorem 4.9.

Once the optimal solution from the first stage is obtained, program (5.3) can be used to calculate the efficiency of the sub-DMU_o in the second stage (follower).

$$e_{2f}^{*(o)}=\text{Max }\mathbb{P}\left(\frac{U\widetilde{y}_o}{W\widetilde{z}_o}\ge\beta_o\right) \tag{5.3}$$

$$\text{s.t.}\quad \mathbb{P}\left(\frac{W\widetilde{z}_j}{V\widetilde{x}_j}\le\beta_j\right)\ge 1-\alpha_j\ (\forall j), \tag{5.3a}$$

$$\mathbb{P}\left(\frac{U\widetilde{y}_j}{W\widetilde{z}_j}\le\beta_j\right)\ge 1-\alpha_j\ (\forall j), \tag{5.3b}$$

$$\mathbb{P}\left(\frac{W\widetilde{z}_o}{V\widetilde{x}_o}\ge\beta_o\right)\ge\Phi\left(e_{1l}^{*(o)}\right), \tag{5.3c}$$

$$W, U, V\ge 0. \tag{5.3d}$$

where $e_{1l}^{*(o)}$ represents the relative efficiency of sub-DMU_o in Stage 1 obtained from program (5.2). As one can see, the objective function and the constraints of the first-stage program are considered as constraints in program (5.3).

In what follows, for computational and operational purposes, we use Cooper et al.'s (1996) approach to compute the deterministic equivalent form of program (5.3).

From constraint (5.3a), we have $\mathbb{P}\left(W\widetilde{z}_j-\beta_j V\widetilde{x}_j\le 0\right)\ge 1-\alpha_j$, which implies that

$$\mathbb{P}\left(\frac{W\widetilde{z}_j-\beta_j V\widetilde{x}_j-\left(Wz_j-\beta_j Vx_j\right)}{\sigma_{W\widetilde{z}_j-\beta_j V\widetilde{x}_j}}\le\frac{-\left(Wz_j-\beta_j Vx_j\right)}{\sigma_{W\widetilde{z}_j-\beta_j V\widetilde{x}_j}}\right)\ge 1-\alpha_j, \tag{5.4}$$

where $z_j=\mathbb{E}\left[\widetilde{z}_j\right]$, $x_j=\mathbb{E}\left[\widetilde{x}_j\right]$, and $\sigma_{W\widetilde{z}_j-\beta_j V\widetilde{x}_j}=\sqrt{Var\left(W\widetilde{z}_j-\beta_j V\widetilde{x}_j\right)}$.

Next, we introduce a new variable $\widetilde{\xi}_j$ as

$$\widetilde{\xi}_j = \frac{W\widetilde{z}_j - \beta_j V\widetilde{x}_j - \left(Wz_j - \beta_j Vx_j\right)}{\sigma_{W\widetilde{z}_j - \beta_j V\widetilde{x}_j}} \ (\forall j), \tag{5.5}$$

which follows the standard normal probability distribution with zero mean and unit variance, *i.e.* $\widetilde{\xi}_j \widetilde{N}(0, 1)$. The direct substitution of $\widetilde{\xi}_j$ in (5.4) yields

$$\mathbb{P}\left(\widetilde{\xi}_j \leq \frac{-\left(Wz_j - \beta_j Vx_j\right)}{\sigma_{W\widetilde{z}_j - \beta_j V\widetilde{x}_j}}\right) \geq 1 - \alpha_j \ (\forall j). \tag{5.6}$$

Since $\widetilde{\xi}_j$ follows the standard normal probability distribution, we have

$$\Phi\left(\frac{-\left(Wz_j - \beta_j Vx_j\right)}{\sigma_{W\widetilde{z}_j - \beta_j V\widetilde{x}_j}}\right) \geq 1 - \alpha_j \ (\forall j). \tag{5.7}$$

According to the property of invertibility of Φ, (5.7) can be rewritten as

$$\frac{-\left(Wz_j - \beta_j Vx_j\right)}{\sigma_{W\widetilde{z}_j - \beta_j V\widetilde{x}_j}} \geq \Phi^{-1}\left(1 - \alpha_j\right) \ (\forall j). \tag{5.8}$$

Now, by introducing the non-negative "spacer variables" η_j, as introduced by Charnes and Cooper (1963), we replace (5.8) with the following two inequalities:

$$\left(Wz_j - \beta_j Vx_j\right) - \Phi^{-1}\left(\alpha_j\right)\eta_j \leq 0 \ (\forall j), \tag{5.9}$$

$$C_{\alpha_j}\left[W^t \Sigma_{zz} W + \beta_j^2 V^t \Sigma_{xx} V - 2\beta_j W^t \Sigma_{zx} V - \eta_j^2\right] \leq 0 \ (\forall j),$$

where

$$C_{\alpha_j} = \begin{cases} 1 & \textit{if } \alpha_j < 0.5, \\ 0 & \textit{if } \alpha_j = 0.5, \\ -1 & \textit{if } \alpha_j > 0.5. \end{cases}$$

Similarly, constraint (5.3b) in program (5.3) could be transformed into the following inequalities:

$$\left(Uy_j - \beta_j Wz_j\right) - \Phi^{-1}\left(\alpha_j\right)\zeta_j \leq 0 \ (\forall j), \tag{5.10}$$

$$C_{\alpha_j}\left[U^t\Sigma_{yy}U + \beta_j^2 W^t\Sigma_{zz}W - 2\beta_j W^t\Sigma_{zx}V - \zeta_j^2\right] \le 0\ (\forall j).$$

For the constraint (5.3c) in (5.3),

$$\mathbb{P}\left(\frac{W\widetilde{z}_o}{V\widetilde{x}_o} \ge \beta_o\right) \ge \Phi\left(e_{1l}^{*(o)}\right)$$

$$\Rightarrow \quad \mathbb{P}(\beta_o V\widetilde{x}_o - W\widetilde{z}_o \le 0) \ge \Phi\left(e_{1l}^{*(o)}\right)$$

$$\Rightarrow \mathbb{P}\left(\frac{(\beta_o V\widetilde{x}_o - W\widetilde{z}_o) - (\beta_o V x_0 - W z_0)}{\sigma_{\beta_o V\widetilde{x}_o - W\widetilde{z}_o}} \le \frac{-(\beta_o V x_0 - W z_0)}{\sigma_{\beta_o V\widetilde{x}_o - W\widetilde{z}_o}}\right)$$

$$\ge \Phi\left(e_{1l}^{*(o)}\right), \tag{5.11}$$

where $z_o = \mathbb{E}[\widetilde{z}_o]$ and $x_o = \mathbb{E}[\widetilde{x}_o]$.

Now, we introduce a new variable $\widetilde{\gamma}_o$ as $\widetilde{\gamma}_0 = \frac{(\beta_o V\widetilde{x}_o - W\widetilde{z}_o) - (\beta_o V x_0 - W z_0)}{\sigma_{\beta_o V\widetilde{x}_0 - W\widetilde{z}_0}}$ so that $\widetilde{\gamma}_o \widetilde{N}(0,1)$. The direct substitution of $\widetilde{\gamma}_o$ in (5.11) yields $\mathbb{P}\left(\widetilde{\gamma}_o \le \frac{-(\beta_o V x_0 - W z_0)}{\sigma_{\beta_o V\widetilde{x}_o - W\widetilde{z}_o}}\right) \ge \Phi\left(e_{1l}^{*(o)}\right)$, which, in turn, implies that $\Phi\left(\frac{-(\beta_o V x_0 - W z_0)}{\sigma_{\beta_o V\widetilde{x}_o - W\widetilde{z}_o}}\right) \ge \Phi\left(e_{1l}^{*(o)}\right)$. Since Φ is a strictly increasing function, $\frac{-(\beta_o V x_0 - W z_0)}{\sigma_{\beta_o V\widetilde{x}_o - W\widetilde{z}_o}} \ge e_{1l}^{*(o)}$.

Using the Charnes and Cooper's (1963) approach, we now introduce a non-negative variable ς_o to convert constraint (5.3c) into the following two inequalities:

$$(W z_o - \beta_o V x_o) - e_{1l}^{*(o)}\ \varsigma_o \le 0, \tag{5.12}$$

$$C_{e_{1l}^{*(o)}}\left[\beta_o^2 V^t\Sigma_{x_o x_o}V + W^t\Sigma_{z_o z_o}W - 2\beta_o W^t\Sigma_{z_o x_o}V - \varsigma_o^2\right] \le 0,$$

where

$$C_{e_{1l}^{*(o)}} = \begin{cases} 1 & \textit{if } e_{1l}^{*(o)} > 0, \\ 0 & \textit{if } e_{1l}^{*(o)} = 0, \\ -1 & \textit{if } e_{1l}^{*(o)} < 0. \end{cases}$$

Now, by replacing (5.3a), (5.3b), and (5.3c) with (5.9), (5.10), and (5.12), respectively, we have

$$
\begin{aligned}
e_{2f}^{*(o)} &= \text{Max } \mathbb{P}\left(\frac{U\widetilde{y}_o}{W\widetilde{z}_o} \geq \beta_o \right) \qquad (5.13)\\
\text{s.t.} \quad & \left(Uy_j - \beta_j Wz_j\right) - \Phi^{-1}\left(\alpha_j\right)\zeta_j \leq 0 \ (\forall j),\\
& C_{\alpha_j}\left[U^t\Sigma_{yy}U + \beta_j^2 W^t\Sigma_{zz}W - 2\beta_j U^t\Sigma_{yz}W - \zeta_j^2\right] \leq 0 \ (\forall j),\\
& \left(Wz_j - \beta_j Vx_j\right) - \Phi^{-1}\left(\alpha_j\right)\xi_j \leq 0 \ (\forall j),\\
& C_{\alpha_j}\left[W^t\Sigma_{zz}W + \beta_j^2 V^t\Sigma_{xx}V - 2\beta_j W^t\Sigma_{zx}V - \xi_j^2\right] \leq 0 \ (\forall j),\\
& \left(Wz_o - \beta_o Vx_o\right) - e_{1l}^{*(o)}\ \varsigma_o \leq 0,\\
& C_{e_{1l}^{*(o)}}\left[\beta_o^2 V^t\Sigma_{x_ox_o}V + W^t\Sigma_{z_oz_o}W - 2\beta_o W^t\Sigma_{z_ox_o}V - \varsigma_o^2\right] \leq 0,\\
& W_d, U_r, V_i, \zeta_j, \xi_j, \varsigma_o \geq 0.
\end{aligned}
$$

Note that all the constraints in program (5.13) are deterministic. However, since the random vectors $\widetilde{y}_o$ and $\widetilde{z}_o$ are included in the objective function, this program is not yet in deterministic form. As it can be seen, program (5.13) can be equivalently set up as

$$
\begin{aligned}
e_{2f}^{*(o)} &= \text{Max } \gamma \qquad (5.14)\\
\text{s.t.} \quad & \mathbb{P}\left(\frac{U\widetilde{y}_o}{W\widetilde{z}_o} \geq \beta_o \right) \geq \gamma,\\
& \left(Uy_j - \beta_j Wz_j\right) - \Phi^{-1}\left(\alpha_j\right)\zeta_j \leq 0 \ (\forall j),\\
& C_{\alpha_j}\left[U^t\Sigma_{yy}U + \beta_j^2 W^t\Sigma_{zz}W - 2\beta_j U^t\Sigma_{yz}W - \zeta_j^2\right] \leq 0 \ (\forall j),\\
& \left(Wz_j - \beta_j Vx_j\right) - \Phi^{-1}\left(\alpha_j\right)\xi_j \leq 0 \ (\forall j),\\
& C_{\alpha_j}\left[W^t\Sigma_{zz}W + \beta_j^2 V^t\Sigma_{xx}V - 2\beta_j W^t\Sigma_{zx}V - \xi_j^2\right] \leq 0 \ (\forall j),\\
& \left(Wz_o - \beta_o Vx_o\right) - e_{1l}^{*(o)}\ \varsigma_o \leq 0,\\
& C_{e_{1l}^{*(o)}}\left[\beta_o^2 V^t\Sigma_{x_ox_o}V + W^t\Sigma_{z_oz_o}W - 2\beta_o W^t\Sigma_{z_ox_o}V - \varsigma_o^2\right] \leq 0,\\
& W_d, U_r, V_i, \zeta_j, \xi_j, \varsigma_o \geq 0.
\end{aligned}
$$

In the spirit of Cooper et al. (1996), the deterministic equivalent form of the first constraint in (5.14) is obtained as follows:

$$\mathbb{P}\left(\frac{U\tilde{y}_o}{W\tilde{z}_o} \geq \beta_o\right) \geq \gamma \ \Rightarrow \mathbb{P}(\, U\tilde{y}_o \geq \beta_o W\tilde{z}_o \,) \geq \gamma$$

$$\Rightarrow \mathbb{P}(\, \beta_o W\tilde{z}_o - U\tilde{y}_o \leq 0 \,) \geq \gamma$$

$$\Rightarrow \mathbb{P}\left(\frac{\beta_o W\tilde{z}_o - U\tilde{y}_o - (\beta_o W z_o - U y_o)}{\sigma_{\beta_o W\tilde{z}_o - U\tilde{y}_o}} \leq \frac{-(\beta_o W z_o - U y_o)}{\sigma_{\beta_o W\tilde{z}_o - U\tilde{y}_o}}\right) \geq \gamma$$

$$\Rightarrow \Phi\left(\frac{U y_o - \beta_o W z_o}{\sigma_{\beta_o W\tilde{z}_o - U\tilde{y}_o}}\right) \geq \gamma$$

$$\Rightarrow \quad \frac{U y_o - \beta_o W z_o}{\sigma_{\beta_o W\tilde{z}_o - U\tilde{y}_o}} \geq \Phi^{-1}(\gamma).$$

Hence, program (5.14) can be expressed as the following program:

$$
\begin{aligned}
e_{2f}^{*(o)} &= \text{Max } \gamma \\
\text{s.t.} \quad & \frac{U y_o - \beta_o W z_o}{\sigma_{\beta_o W\tilde{z}_o - U\tilde{y}_o}} \geq \Phi^{-1}(\gamma), \\
& \left(U y_j - \beta_j W z_j\right) - \Phi^{-1}(\alpha_j)\zeta_j \leq 0 \ (\forall j), \\
& C_{\alpha_j}\left[U^t \Sigma_{yy} U + \beta_j^2 W^t \Sigma_{zz} W - 2\beta_j U^t \Sigma_{yz} W - \zeta_j^2\right] \leq 0 \ (\forall j), \\
& \left(W z_j - \beta_j V x_j\right) - \Phi^{-1}(\alpha_j)\xi_j \leq 0 \ (\forall j), \\
& C_{\alpha_j}\left[W^t \Sigma_{zz} W + \beta_j^2 V^t \Sigma_{xx} V - 2\beta_j W^t \Sigma_{zx} V - \xi_j^2\right] \leq 0 \ (\forall j), \\
& (W z_o - \beta_o V x_o) - e_{1l}^{*(o)} \varsigma_o \leq 0, \\
& C_{e_{1l}^{*(o)}}\left[\beta_o^2 V^t \Sigma_{x_o x_o} V + W^t \Sigma_{z_o z_o} W - 2\beta_o W^t \Sigma_{z_o x_o} V - \varsigma_o^2\right] \leq 0, \\
& W_d, U_r, V_i, \zeta_j, \xi_j, \varsigma_o \geq 0.
\end{aligned}
\tag{5.15}
$$

Although program (5.15) is deterministic, it is not easy to solve. This is because the denominator in the first constraint makes this program a non-convex programming problem (Cooper et al., 1996). To circumvent this difficulty, consider the following program:

$$
\begin{aligned}
e_{2f}^{*(o)} &= \text{Max } \theta \\
\text{s.t.} \quad & \frac{U y_o - \beta_o W z_o}{\sigma_{\beta_o W\tilde{z}_o - U\tilde{y}_o}} \geq \theta,
\end{aligned}
\tag{5.16}
$$

$$\left(Uy_j - \beta_j Wz_j\right) - \Phi^{-1}\left(\alpha_j\right)\zeta_j \leq 0 \ (\forall j),$$

$$C_{\alpha_j}\left[U^t \Sigma_{yy} U + \beta_j W^t \Sigma_{zz} W - 2\beta_j U^t \Sigma_{yz} W - \zeta_j^2\right] \leq 0 \ (\forall j),$$

$$\left(Wz_j - \beta_j Vx_j\right) - \Phi^{-1}\left(\alpha_j\right)\xi_j \leq 0 \ (\forall j),$$

$$C_{\alpha_j}\left[W^t \Sigma_{zz} W + \beta_j^2 V^t \Sigma_{xx} V - 2\beta_j W^t \Sigma_{zx} V - \xi_j^2\right] \leq 0 \ (\forall j),$$

$$\left(Wz_o - \beta_o Vx_o\right) - e_{1l}^{*(o)} \varsigma_o \leq 0 \ (\forall j),$$

$$C_{e_{1l}^{*(o)}}\left[\beta_o^2 V^t \Sigma_{x_o x_o} V + W^t \Sigma_{z_o z_o} W - 2\beta_o W^t \Sigma_{z_o x_o} V - \varsigma_o^2\right] \leq 0 \ (\forall j),$$

$$W_d, U_r, V_i, \zeta_j, \xi_j, \varsigma_o \geq 0.$$

Since $\Phi^{-1}(\gamma)$ in the first constraint of (5.15) is a strictly increasing function of γ, programs (5.15) and (5.16) have the same optimal solution, *i.e.*

$$\theta^* = \Phi^{-1}(\gamma^*). \tag{5.17}$$

As it can be seen, program (5.16) is equivalent to the following problem:

$$e_{2f}^{*(o)} = \text{Max} \ \frac{Uy_o - \beta_o Wz_o}{\sigma_{\beta_o W\widetilde{z}_o - U\widetilde{y}_o}} \tag{5.18}$$

$$\text{s.t.} \quad \left(Uy_j - \beta_j Wz_j\right) - \Phi^{-1}\left(\alpha_j\right)\zeta_j \leq 0 \ (\forall j),$$

$$C_{\alpha_j}\left[U^t \Sigma_{yy} U + \beta_j^2 W^t \Sigma_{zz} W - 2\beta_j U^t \Sigma_{yz} W - \zeta_j^2\right] \leq 0 \ (\forall j),$$

$$\left(Wz_j - \beta_j Vx_j\right) - \Phi^{-1}\left(\alpha_j\right)\xi_j \leq 0 \ (\forall j),$$

$$C_{\alpha_j}\left[W^t \Sigma_{zz} W + \beta_j^2 V^t \Sigma_{xx} V - 2\beta_j W^t \Sigma_{zx} V - \xi_j^2\right] \leq 0 \ (\forall j),$$

$$\left(Wz_o - \beta_o Vx_o\right) - e_{1l}^{*(o)} \varsigma_o \leq 0,$$

$$C_{e_{1l}^{*(o)}}\left[\beta_o^2 V^t \Sigma_{x_o x_o} V + W^t \Sigma_{z_o z_o} W - 2\beta_o W^t \Sigma_{z_o x_o} V - \varsigma_o^2\right] \leq 0,$$

$$W_d, U_r, V_i, \zeta_j, \xi_j, \varsigma_o \geq 0.$$

Since $(Uy_o - \beta_o Wz_o)/\sigma_{\beta_o W\widetilde{z}_o - U\widetilde{y}_o}$ is bounded from above by $\Phi^{-1}(\alpha_o)$, we introduce a positive variable ϖ to convert program (5.18) into the following program:

$$e_{2f}^{*(o)} = \text{Max} \ \frac{Uy_o - \beta_o Wz_o}{\varpi} \tag{5.19}$$

$$\begin{aligned}
\text{s.t.} \quad & \beta_o^2\, W^t \Sigma_{z_o z_o} W + U^t \Sigma_{y_o y_o} U - 2\beta_o U^t \Sigma_{y_o z_o} W - \varpi^2 \geq 0, \\
& \left(U y_j - \beta_j W z_j\right) - \Phi^{-1}(\alpha_j)\zeta_j \leq 0 \ (\forall j), \\
& C_{\alpha_j}\left[U^t \Sigma_{yy} U + \beta_j^2 W^t \Sigma_{zz} W - 2\beta_j U^t \Sigma_{yz} W - \zeta_j^2\right] \leq 0 \ (\forall j), \\
& \left(W z_j - \beta_j V x_j\right) - \Phi^{-1}(\alpha_j)\xi_j \leq 0 \ (\forall j), \\
& C_{\alpha_j}\left[W^t \Sigma_{zz} W + \beta_j^2 V^t \Sigma_{xx} V - 2\beta_j W^t \Sigma_{zx} V - \xi_j^2\right] \leq 0 \ (\forall j), \\
& (W z_o - \beta_o V x_o) - e_{1l}^{*(o)} \varsigma_o \leq 0, \\
& C_{e_{1l}^{*(o)}}\left[\beta_o^2 V^t \Sigma_{x_o x_o} V + W^t \Sigma_{z_o z_o} W - 2\beta_o W^t \Sigma_{z_o x_o} V - \varsigma_o^2\right] \leq 0, \\
& W_d, U_r, V_i, \zeta_j, \xi_j, \varsigma_o \geq 0.
\end{aligned}$$

Since program (5.19) involves a fractional objective function, we use Charnes and Cooper's (1962) transformation, to convert this into the following quadratic programming problem:

$$e_{2f}^{*(o)} = \text{Max}\ \mu y_o - \beta_o \lambda z_o \tag{5.20}$$

$$\begin{aligned}
\text{s.t.} \quad & \beta_o^2\, \lambda^t \Sigma_{z_o z_o} \lambda + \mu^t \Sigma_{y_o y_o} \mu - 2\beta_o \mu^t \Sigma_{y_o z_o} \lambda \geq 1, \\
& \left(\mu y_j - \beta_j \lambda z_j\right) - \Phi^{-1}(\alpha_j)\tau_j \leq 0 \ (\forall j), \\
& C_{\alpha_j}\left[\mu^t \Sigma_{yy} \mu + \beta_j^2 \lambda^t \Sigma_{zz} \lambda - 2\beta_j \mu^t \Sigma_{yz} \lambda - \tau_j^2\right] \leq 0 \ (\forall j), \\
& \left(\lambda z_j - \beta_j \upsilon x_j\right) - \Phi^{-1}(\alpha_j)\rho_j \leq 0 \ (\forall j), \\
& C_{\alpha_j}\left[\lambda \Sigma_{zz} \lambda + \beta_j^2 \upsilon^t \Sigma_{xx} \upsilon - 2\beta_j \lambda^t \Sigma_{zx} \upsilon - \rho_j^2\right] \leq 0 \ (\forall j), \\
& (\lambda z_o - \beta_o \upsilon x_o) - e_{1l}^{*(o)} \pi_o \leq 0, \\
& C_{e_{1l}^{*(o)}}\left[\beta_o^2 \upsilon^t \Sigma_{x_o x_o} \upsilon + \lambda^t \Sigma_{z_o z_o} \lambda - 2\beta_o \lambda^t \Sigma_{z_o x_o} \upsilon - \pi_o^2\right] \leq 0, \\
& \lambda_d, \mu_r, \upsilon_i, \tau_j, \rho_j, \pi_o \geq 0,
\end{aligned}$$

where $t := \frac{1}{\varpi}, \lambda := tW, \ \mu := tU, \ \upsilon := tV, \ \tau_j := t\zeta_j, \ \rho_j := t\xi_j, \ \pi_o := t\varsigma_o$.

Now, to conclude this section, we present Theorem 5.2 below to show the relationship between the solutions obtained from programs (5.3) and (5.20).

Theorem 5.2 Suppose $\beta_{jo} = \beta_o = 1$ ($\beta_{jo} = \beta_o < 1$). Let $(\lambda^*, \mu^*, v^*, \tau^*, \rho^*, \pi^*)$ and (W^*, U^*, V^*) be the optimal solutions obtained from programs (5.20) and (5.3), respectively. Then, $\Phi(\mu^* y_o - \beta_o \lambda^* z_o) = \mathbb{P}\left(\frac{U^* \widetilde{y_o}}{W^* \widetilde{z_o}} \geq \beta_o\right)$. Furthermore, DMU_o is stochastically (satisficing) efficient if and only if $\Phi(\mu^* y_o - \beta_o \lambda^* z_o) = \alpha_0$.

Proof From (5.17), we have $\Phi(\mu^* y_o - \beta_o \lambda^* z_o) = \gamma^*$, where γ^* is the optimal value obtained from program (5.15). Since program (5.15) is equivalent to program (5.3), then $\gamma^* = \mathbb{P}\left(\frac{U^* \widetilde{y}_o}{W^* \widetilde{z}_o} \geq \beta_o\right)$ and, consequently, $\Phi(\mu^* y_o - \beta_o \lambda^* z_o) = \mathbb{P}\left(\frac{U^* \widetilde{y}_o}{W^* \widetilde{z}_o} \geq \beta_o\right)$. ∎

5.2.2 Transformation to Linear Programming

Following Cooper et al. (1996), we adopt the single-factor symmetric disturbance assumption wherein all the components of input, intermediate measure, and output are determined by a single random factor. Sharpe (1963) and Kahane (1977) examined the applications of the single underlying factor assumption in finance and economics. More explicitly, suppose that $\widetilde{x}_{ij} = x_{ij} + a_{ij}\,\xi$, $\widetilde{z}_{dj} = z_{dj} + b_{dj}\,\eta$, and $\widetilde{y}_{rj} = y_{rj} + c_{rj}\,\epsilon$, where x_{ij}, z_{dj}, y_{rj}, a_{ij}, b_{dj},and c_{rj} are non-negative and ξ, η,and ϵ are the independent standard normal distributed random variables. Also, x_{ij}, z_{dj}, and y_{rj} are assumed to be the expected values for $\widetilde{x}_{ij}$, $\widetilde{z}_{dj}$, and $\widetilde{y}_{rj}$, respectively, and a_{ij}, b_{dj}, and c_{rj} are the standard deviations for $\widetilde{x}_{ij}, \widetilde{z}_{dj}$, and $\widetilde{y}_{rj}$. Meanwhile, having finite realizations and non-negative distributions are reasonable restrictions for all $\widetilde{x}_{ij}$, $\widetilde{z}_{dj}$, and $\widetilde{y}_{rj}$ (Zhou et al., 2017). Like the analysis performed in the previous section, program (5.1) can be transformed as

$$\begin{aligned}
e_{1l}^{*(o)} &= \text{Max } (wz_0 - \beta_o vx_0) \qquad (5.21)\\
\text{s.t.} \quad & |\beta_o va_o - wb_o| \geq 1,\\
(\beta_j vx_j - wz_j) &\geq \Phi^{-1}(1-\alpha_j)\left|\beta_j va_j - wb_j\right| \; (\forall j),\\
& w, v \geq 0.
\end{aligned}$$

Since the first constraint of program (5.21) includes an absolute function, program (5.21) is non-linear. However, using the goal programming theory by Charnes and Cooper (1961, 1977), program (5.21) can be transformed into a quadratic programming problem (Cooper et al., 1996). Suppose that

$$\begin{aligned}
\left|\beta_j va_j - wb_j\right| &= p_j^+ + p_j^-,\\
\beta_j va_j - wb_j &= p_j^+ - p_j^-,\\
p_j^+ p_j^- &= 0,\\
p_j^+, p_j^- &\geq 0.
\end{aligned}$$

Using the above, program (5.21) can be transformed into the following form:

$$\begin{aligned} e_{1l}^{*(o)} &= \text{Max } (wz_0 - \beta_o vx_0) \quad (5.22)\\ \text{s.t.} \quad & p_o^+ + p_o^- \geq 1,\\ (\beta_o vx_j - wz_j) &\geq \Phi^{-1}(1-\alpha_j)\left(p_j^+ + p_j^-\right) \ (\forall j),\\ \beta_j va_j - wb_j &= p_j^+ - p_j^- \ (\forall j),\\ p_j^+ p_j^- &= 0 \ (\forall j),\\ w, v, p^+, p^- &\geq 0. \end{aligned}$$

Due to the existence of the constraint $p_j^+ p_j^- = 0$, program (5.22) is non-linear. If we ignore this constraint, the program becomes linear. Moreover, as we know, if the linear programming problem has an optimal solution, then it has an optimal extreme point and at least one of the variables p_j^+ *or* p_j^- is zero. Consequently, if we use the simplex algorithm to solve this linear programming problem, we can find this optimal extreme point and can hence avoid the constraint $p_j^+ p_j^- = 0$ without any loss of any information. Therefore, we just need to solve the following linear program using the simplex algorithm:

$$\begin{aligned} e_{1l}^{*(o)} &= \text{Max } (wz_0 - \beta_o vx_0) \quad (5.23)\\ \text{s.t.} \quad & p_o^+ + p_o^- \geq 1,\\ (\beta_j vx_j - wz_j) &\geq \Phi^{-1}(1-\alpha_j)\left(p_j^+ + p_j^-\right) \ (\forall j),\\ \beta_j va_j - wb_j &= p_j^+ - p_j^- \ (\forall j),\\ w, v, p^+, p^- &\geq 0. \end{aligned}$$

Also, for the follower in Stage 2, like the analysis performed in the previous section, program (5.3) can be transformed as

$$\begin{aligned} e_{2f}^{*(o)} &= \text{Max } uy_o - \beta_o wz_o \quad (5.24)\\ \text{s.t.} \quad & |\beta_o wb_o - uc_o| \geq 1,\\ (\beta_j wz_j - uy_j) &\geq \Phi^{-1}(1-\alpha_j)|\beta_j wb_j - uc_j| \ (\forall j),\\ (\beta_j vx_j - wz_j) &\geq \Phi^{-1}(1-\alpha_j)|\beta_j va_j - wb_j| \ (\forall j),\\ (wz_o - \beta_o vx_o) &\geq e_{1l}^{*(o)}|\beta_o va_o - wb_o|,\\ v, w, u &\geq 0. \end{aligned}$$

Now, by maintaining the following assumptions:

$$\left|\beta_j w b_j - u c_j\right| = q_j^+ + q_j^-,$$

$$\beta_j w b_j - u c_j = q_j^+ - q_j^-,$$

$$q_j^+ q_j^- = 0,$$

$$q_j^+, q_j^- \geq 0.$$

$$\left|\beta_j v a_j - w b_j\right| = p_j^+ + p_j^-,$$

$$\beta_j v a_j - w b_j = p_j^+ - p_j^-,$$

$$p_j^+ p_j^- = 0,$$

$$p_j^+, p_j^- \geq 0,$$

and by replacing them in program (5.24), we obtain the following non-linear program:

$$
\begin{aligned}
e_{2f}^{*(o)} &= \text{Max } u y_o - \beta_o w z_o \\
\text{s.t.} \quad & q_o^+ + q_o^- \geq 1, \\
& \left(\beta_j w z_j - u y_j\right) \geq \Phi^{-1}\left(1 - \alpha_j\right)\left(q_j^+ + q_j^-\right) \ (\forall j), \\
& \beta_j w b_j - u c_j = q_j^+ - q_j^- \ (\forall j), \\
& \left(\beta_j \upsilon x_j - w z_j\right) \geq \Phi^{-1}\left(1 - \alpha_j\right)\left(p_j^+ + p_j^-\right) \ (\forall j), \\
& \beta_j v a_j - w b_j = p_j^+ - p_j^- \ (\forall j), \\
& \left(w z_o - \beta_o v x_o\right) \geq e_{1l}^{*(o)}\left(p_o^+ + p_o^-\right), \\
& q_j^+ q_j^- = 0 \ (\forall j), \\
& p_j^+ p_j^- = 0 \ (\forall j), \\
& u, w, v, p^+, p^-, q^+, q^- \geq 0.
\end{aligned}
\tag{5.25}
$$

Like what was done before, the linear form of program (5.25) can be set up as

$$
\begin{aligned}
e_{2f}^{*(o)} &= \text{Max } u y_o - \beta_o w z_o \\
\text{s.t.} \quad & q_o^+ + q_o^- \geq 1, \\
& \left(\beta_j w z_j - u y_j\right) \geq \Phi^{-1}\left(1 - \alpha_j\right)\left(q_j^+ + q_j^-\right) \ (\forall j),
\end{aligned}
\tag{5.26}
$$

$$\beta_j wb_j - uc_j = q_j^+ - q_j^- \quad (\forall j),$$

$$(\beta_j vx_j - wz_j) \geq \Phi^{-1}(1-\alpha_j)\left(p_j^+ + p_j^-\right) \quad (\forall j),$$

$$\beta_j va_j - wb_j = p_j^+ - p_j^- \quad (\forall j),$$

$$(wz_o - \beta_o vx_o) \geq e_{1l}^{*(o)}(p_o^+ + p_o^-),$$

$$u, w, v, p^+, p^-, q^+, q^- \geq 0.$$

5.2.3 *Stage 2 as the Leader and Stage 1 as the Follower*

Now, suppose that Stage 2 is the leader. In this case, we know the efficiency of the leader based on which the efficiency of the follower can be computed. We set up program (5.27) to compute the efficiency of the sub-DMU$_o$ in Stage 2 (leader).

$$e_{2l}^{*(o)} = \text{Max } \mathbb{P}\left(\frac{U\tilde{y}_o}{W\tilde{z}_o} \geq \beta_o\right) \tag{5.27}$$

$$\text{s.t.} \quad \mathbb{P}\left(\frac{U\tilde{y}_j}{W\tilde{z}_j} \leq \beta_j\right) \geq 1 - \alpha_j \ (\forall j),$$

$$W, U \geq 0.$$

Using Cooper et al.'s (1996) transformation in stochastic programming, the deterministic form of program (5.27) can be set up as

$$e_{2l}^{*(o)} = \text{Max } (uy_0 - \beta_o wz_0) \tag{5.28}$$

$$\text{s.t.} \quad \beta_o^2 w^t \Sigma_{z_0 z_o} w + u^t \Sigma_{y_0 y_0} u - 2\beta_o w^t \Sigma_{z_0 y_0} u \geq 1,$$

$$\left(uy_j - \beta_j wz_j\right) - \Phi^{-1}(\alpha_j)\xi_j \leq 0 \ (\forall j),$$

$$C_{\alpha_j}\left[\beta_j^2 w^t \Sigma_{zz} w + u^t \Sigma_{yy} u - 2\beta_j w^t \Sigma_{zy} u - \xi_j^2\right] \leq 0 \ (\forall j),$$

$$w, u, \xi \geq 0.$$

The relationship between the optimal solutions obtained from programs (5.27) and (5.28) is presented in Theorem 5.3.

Theorem 5.3 Suppose $\beta_{jo} = \beta_o = 1$ ($\beta_{jo} = \beta_o < 1$) and let (W^*, U^*) and (w^*, u^*, ξ^*) be the optimal solutions obtained from programs (5.27) and (5.28), respectively.

Then, $\Phi(u^* y_0 - \beta_o w^* z_0) = \mathbb{P}\left(\frac{U^* \widetilde{y_o}}{W^* \widetilde{z_o}} \geq \beta_o\right)$. Furthermore, DMU_o is stochastically (satisficing) efficient if and only if $\Phi(u^* y_0 - \beta_o w^* z_0) = \alpha_0$.

Besides, by assuming single-factor components of any input, intermediate measure, and output, the deterministic linear programming model when $\alpha \leq 0.5$ could be obtained as follows:

$$
\begin{aligned}
e_{2l}^{*(o)} &= \text{Max } (u y_0 - \beta_o w z_0) \\
\text{s.t.} \quad & | \beta_o w b_o - u c_o | \geq 1, \\
(\beta_j w z_j - u y_j) &\geq \Phi^{-1}(1 - \alpha_j) \left| \beta_j w b_j - u c_j \right| \ (\forall j), \\
& w, v \geq 0.
\end{aligned}
\tag{5.29}
$$

Since the first constraint of program (5.29) includes an absolute form, this program is non-linear. However, using the goal programming theory by Charnes and Cooper (1961, 1977), program (5.29) can be transformed into a quadratic programming problem (Cooper et al., 1996). Again, by assuming the following:

$$
\begin{aligned}
\left| \beta_j w b_j - u c_j \right| &= q_j^+ + q_j^-, \\
\beta_j w b_j - u c_j &= q_j^+ - q_j^-, \\
q_j^+ q_j^- &= 0, \\
q_j^+, q_j^- &\geq 0,
\end{aligned}
$$

and replacing them in program (5.29), we obtain the following program:

$$
\begin{aligned}
e_{2l}^{*(o)} &= \text{Max } (u y_0 - \beta_o w z_0) \\
\text{s.t.} \quad & q_o^+ + q_o^- \geq 1, \\
(\beta_o w z_j - u y_j) &\geq \Phi^{-1}(1 - \alpha_j) \left(q_j^+ + q_j^- \right) \ (\forall j), \\
\beta_j w b_j - u c_j &= q_j^+ - q_j^- \ (\forall j), \\
q_j^+ q_j^- &= 0 \ (\forall j), \\
& w, u, q^+, q^- \geq 0.
\end{aligned}
\tag{5.30}
$$

If we ignore constraint $q_j^+ q_j^- = 0$, the linear form of program (5.30) is

$$
e_{2l}^{*(o)} = \text{Max } (u y_0 - \beta_o w z_0) \tag{5.31}
$$

$$\text{s.t.} \quad q_o^+ + q_o^- \geq 1,$$
$$\left(\beta_j w z_j - u y_j\right) \geq \Phi^{-1}(1-\alpha_j)\left(q_j^+ + q_j^-\right) \ (\forall j),$$
$$\beta_j w b_j - u c_j = q_j^+ - q_j^- \ (\forall j),$$
$$w, u, q^+, q^- \geq 0.$$

Once the optimal solution of program (5.31) is obtained, we set up the following program to compute the relative efficiency of sub-DMU_o in Stage 1 (follower):

$$e_{1\text{f}}^{*(o)} = \text{Max } \mathbb{P}\left(\frac{W\tilde{z}_o}{V\tilde{x}_o} \geq \beta_o\right) \tag{5.32}$$
$$\text{s.t.} \quad \mathbb{P}\left(\frac{W\tilde{z}_j}{V\tilde{x}_j} \leq \upbeta_\text{j}\right) \geq 1-\alpha_j \ (\forall j),$$
$$\mathbb{P}\left(\frac{U\tilde{y}_j}{W\tilde{z}_j} \leq \upbeta_\text{j}\right) \geq 1-\alpha_j \ (\forall j),$$
$$\mathbb{P}\left(\frac{U\tilde{y}_o}{W\tilde{z}_o} \geq \upbeta_\text{o}\right) \geq \Phi\left(e_{21}^{*(o)}\right)$$
$$W, U, V \geq 0,$$

The equivalent deterministic form of program (5.32) can be set up as

$$e_{1f}^{*(o)} = Max \quad wz_o - \beta_o v x_o \tag{5.33}$$
$$s.t. \quad w^t\Sigma_{z_o z_o} w + \beta_o^2 v^t \Sigma_{x_o x_o} v - 2\beta_o w^t \Sigma_{x_o z_o} v \geq 1,$$
$$\left(wz_j - \upbeta_\text{j} v x_j\right) - \Phi^{-1}(\alpha_j)\tau_j \leq 0 \ (\forall j),$$
$$C_{\alpha_j}\left[\upbeta_\text{j}^2 v^t \Sigma_{xx} v + w^t \Sigma_{zz} w - 2\upbeta_\text{j} v^t \Sigma_{xz} w - \tau_j^2\right] \leq 0 \ (\forall j),$$
$$\left(uy_j - \upbeta_\text{j} w z_j\right) - \Phi^{-1}(\alpha_j)\rho_j \leq 0 \ (\forall j),$$
$$C_{\alpha_j}\left[\upbeta_\text{j}^2 w \Sigma_{zz} w + u^t \Sigma_{yy} u - 2\upbeta_\text{j} w^t \Sigma_{zy} u - \rho_j^2\right] \leq 0 \ (\forall j),$$
$$\left(uy_o - \upbeta_\text{o} w z_o\right) - e_{2l}^{*(o)}\pi_o \leq 0,$$
$$C_{e_{2l}^{*(o)}}\left[u^t \Sigma_{y_o y_o} u + \upbeta_\text{o}^2 w^t \Sigma_{z_o z_o} w - 2\upbeta_\text{o} w^t \Sigma_{z_o y_o} u - \pi_o^2\right] \leq 0,$$
$$w_d, u_r, v_i, \tau_j, \rho_j, \pi_o \geq 0,$$

where

$$C_{e_{2l}^{*(o)}} = \begin{cases} 1 & \text{if } e_{2l}^{*(o)} > 0, \\ 0 & \text{if } e_{2l}^{*(o)} = 0, \\ -1 & \text{if } e_{2l}^{*(o)} < 0. \end{cases}$$

Theorem 5.4 Suppose that $\beta_{jo} = \beta_o = 1$ ($\beta_{jo} = \beta_o < 1$) and let (W^*, U^*, V^*) and $(w^*, u^*, v^*, \tau^*, \rho^*, \pi^*)$ be the optimal solutions obtained from programs (5.32) and (5.33), respectively. Then, $\Phi(w^* z_o - \beta_o v^* x_o) = \mathbb{P}\left(\frac{W^* \tilde{z}_o}{V^* \tilde{x}_o} \geq \beta_o\right)$. Moreover, DMU_o is stochastically (satisficing) efficient if and only if $\Phi(w^* z_o - \beta_o v^* x_o) = \alpha_0$.

Finally, using the single-factor assumption, the deterministic form of the linear program for $\alpha \leq 0.5$ can be obtained as

$$\begin{aligned} e_{1f}^{*(o)} &= \text{Max } wz_o - \beta_o vx_o \qquad (5.34) \\ \text{s.t.} \quad & |\beta_o va_o - wb_o| \geq 1, \\ (\beta_j wz_j - uy_j) &\geq \Phi^{-1}(1-\alpha_j)|\beta_j wb_j - uc_j| \quad (\forall j), \\ (\beta_j vx_j - wz_j) &\geq \Phi^{-1}(1-\alpha_j)|\beta_j va_j - wb_j| \quad (\forall j), \\ (uy_0 - \beta_o wz_0) &\geq e_{2l}^{*(o)} |\beta_o wb_o - uc_o|, \\ & v, w, u \geq 0. \end{aligned}$$

Now, by using the following assumptions:

$$\begin{aligned} |\beta_j wb_j - uc_j| &= q_j^+ + q_j^-, \\ \beta_j wb_j - uc_j &= q_j^+ - q_j^-, \\ q_j^+ q_j^- &= 0, \\ q_j^+, q_j^- &\geq 0, \\ |\beta_j va_j - wb_j| &= p_j^+ + p_j^-, \\ \beta_j va_j - wb_j &= p_j^+ - p_j^-, \\ p_j^+ p_j^- &= 0, \\ p_j^+, p_j^- &\geq 0, \end{aligned}$$

program (5.34) can be transformed as

$$
\begin{aligned}
e_{1f}^{*(o)} &= \text{Max } wz_o - \beta_o vx_o \qquad (5.35)\\
\text{s.t.} \quad & p_o^+ + p_o^- \geq 1,\\
(\beta_j wz_j - uy_j) &\geq \Phi^{-1}(1-\alpha_j)\left(q_j^+ + q_j^-\right) \ (\forall j),\\
\beta_j wb_j - uc_j &= q_j^+ - q_j^- \ (\forall j),\\
(\beta_j vx_j - wz_j) &\geq \Phi^{-1}(1-\alpha_j)\left(p_j^+ + p_j^-\right) \ (\forall j),\\
\beta_j va_j - wb_j &= p_j^+ - p_j^- \ (\forall j)\\
(uy_0 - \beta_o wz_0) &\geq e_{2l}^{*(o)}\left(q_o^+ + q_o^-\right),\\
q_j^+ q_j^- &= 0 \ (\forall j),\\
p_j^+ p_j^- &= 0 \ (\forall j),\\
u, w, v, p^+, p^-, q^+, q^- &\geq 0.
\end{aligned}
$$

Finally, the linear form of program (5.35) is set up as

$$
\begin{aligned}
e_{1f}^{*(o)} &= \text{Max } wz_o - \beta_o vx_o \qquad (5.36)\\
\text{s.t.} \quad & p_o^+ + p_o^- \geq 1,\\
(\beta_j wz_j - uy_j) &\geq \Phi^{-1}(1-\alpha_j)\left(q_j^+ + q_j^-\right) \ (\forall j),\\
\beta_j wb_j - uc_j &= q_j^+ - q_j^- \ (\forall j),\\
(\beta_j vx_j - wz_j) &\geq \Phi^{-1}(1-\alpha_j)\left(p_j^+ + p_j^-\right) \ (\forall j),\\
\beta_j va_j - wb_j &= p_j^+ - p_j^- \ (\forall j),\\
(uy_0 - \beta_o wz_0) &\geq e_{2l}^{*(o)}\left(q_o^+ + q_o^-\right),\\
u, w, v, p^+, p^-, q^+, q^- &\geq 0.
\end{aligned}
$$

We now present Theorem (5.5) below to compare the relative efficiency achieved for Stage 1 using programs (5.23) and (5.36).

Theorem 5.5 For every optimal solution obtained from programs (5.23) and (5.36), it holds that $e_{1f}^{*(o)} \leq e_{1l}^{*(o)}$.

Proof Suppose that $(u^*, w^*, v^*, p^{+*}, p^{-*}, q^{+*}, q^{-*})$ is the optimal solution obtained from program (5.36). It is easy to see that this solution is also feasible in program (5.23). Therefore, $e_{1f}^{*(o)} \leq e_{1l}^{*(o)}$. ∎

Theorem 5.6 For every optimal solution obtained from programs (5.26) and (5.31), it holds that $e_{2f}^{*(o)} \leq e_{2l}^{*(o)}$.

Proof The proof is similar to that of Theorem 5.5. ∎

Corollary 5.1. If the sub-DMU$_o$ in Stage 1 (as the follower) is efficient at the confidence level α, then it is also efficient in Stage 2 (as the leader) at the same confidence level α. Also, at the confidence level α, if the sub-DMU$_o$ at each stage is inefficient as the leader, then it is also inefficient as the follower.

Theorem 5.7 For each confidence level $\alpha \leq \alpha'$, $e_{1f}^{*(o)}(\alpha) \geq e_{1f}^{*(o)}(\alpha')$.

Proof Let $(u^*, w^*, v^*, p^{+*}, p^{-*}, q^{+*}, q^{-*})$ be an optimal solution to program (5.36) when we evaluate the performance of DMU$_o$ at the confidence level α. Since $\Phi^{-1}(\alpha)$ is a strictly increasing function, if $\alpha \leq \alpha'$, then $\Phi^{-1}(\alpha) \leq \Phi^{-1}(\alpha')$. Therefore, $(\lambda^* z_j - \beta_j v^* x_j) - \Phi^{-1}(\alpha')\tau_j^* \leq (\lambda^* z_j - \beta_j v^* x_j) - \Phi^{-1}(\alpha)\tau_j^* \leq 0$ and $(\mu^* y_j - \beta_j \lambda^* z_j) - \Phi^{-1}(\alpha')\rho_j^* \leq (\mu^* y_j - \beta_j \lambda^* z_j) - \Phi^{-1}(\alpha)\rho_j^* \leq 0$. Hence, $(u^*, w^*, v^*, p^{+*}, p^{-*}, q^{+*}, q^{-*})$ is also an optimal solution to program (5.36) to evaluate the performance of DMU$_o$ at the confidence level α'. So, we must have $e_{1f}^{*(o)}(\alpha) \geq e_{1f}^{*(o)}(\alpha')$. ∎

Theorem 5.8 For each confidence level $\alpha < \alpha'$, $e_{11}^{*(o)}(\alpha) \geq e_{11}^{*(o)}(\alpha')$, $e_{21}^{*(o)}(\alpha) \geq e_{21}^{*(o)}(\alpha')$, and $e_{2f}^{*(o)}(\alpha) \geq e_{2f}^{*(o)}(\alpha')$.

Corollary 5.2. If in one stage, as leader or follower, the sub-DMU$_o$ is stochastically efficient under the confidence level α' ($\alpha' > \alpha$), then it is stochastically efficient under the confidence level α. Besides, if in one stage, as leader or follower, the sub-DMU$_o$ is not stochastically efficient under the confidence level α, then it is also not stochastically efficient under the confidence level α'.

5.2.4 An Illustrative Application

We now proceed to apply our proposed model to a real case. In performance evaluation studies, the computation of efficiency of commercial banks has been an interesting avenue for research (Seiford & Zhu, 1999; Yang et al., 2011; Paradi & Zhu, 2013; Zhou et al., 2017). Most of the existing studies on performance analysis in the banking industry have considered input and output data that are deterministic. We have considered a data set of 16 commercial banks for 10 years (2000-2010), which was used earlier by Zhou et al. (2017) to illustrate the applicability of the model. This data set deals with the two-stage production process of banks. Stage 1 uses three inputs, namely employees, fixed assets, and expenses to produce two intermediate outputs, namely deposits and interbank deposits, which are then used as inputs in Stage 2 to produce two final outputs, namely loan and profit. All inputs and outputs are assumed to follow a normal distribution. Table 5.1 provides the approximated means and standard deviations for these 16 banks.

Table 5.2 shows the results of overall efficiency scores obtained from the stochastic CCR-DEA model under the aspiration level of $\alpha = 1$ and the confidence levels of $\alpha = 0.1, \ldots, 0.5$, which only considers the inputs of Stage 1 and the final outputs of Stage 2 but ignores the intermediate outputs. As it can be seen, DMU 15 has the highest performance at all levels.

The efficiency scores obtained from this black-box production system can be misleading and not realistic as this system ignores the internal structure of banks by not addressing the potential conflict between the two stages arising from the presence of intermediate outputs.

We now turn to the case where the DMU under evaluation is considered the leader in Stage 1 and the follower in Stage 2. The efficiency scores obtained from programs (5.23) and (5.26) under the aspiration level of $\beta = 1$ and the confidence levels of $\alpha = 0.1, \ldots, 0.5$ are presented in Table 5.3. As it can be seen, DMU 15 exhibits the best performance at $\alpha = 0.1$, implying that it is stochastically efficient in both stages. DMUs 2, 3, 4, 6, 7, 9, 11, and 14 are the worst performers at all the confidence levels considered, implying that they are stochastically inefficient in both stages. Furthermore, DMU 16 is stochastically inefficient in Stage 1 but stochastically efficient in Stage 2, at all the confidence levels considered. On the contrary, DMUs 1 and 13 are stochastically efficient in Stage 1 but stochastically inefficient in Stage 2, at all the confidence levels. By comparing the efficiencies of the DMUs at various confidence levels, one can find out the importance of the confidence levels in judging their performance.

We now proceed to the case where the DMU under evaluation is considered the leader in Stage 2 and the follower in Stage 1. The efficiency scores obtained from programs (5.31) and (5.36) are presented in Table 5.4. DMU 15 is found to exhibit the best performance at all the confidence levels, implying that it is stochastically efficient in both of its stages. And, DMUs 2, 3, 4, 5, 6, 7, 8, 9, 11, 14, and 16 are found to be the worst performers at all the confidence levels, implying that they all have the potential to improve their performance in both the stages.

On comparing the results from Tables 5.3 and 5.4, one can observe that at any fixed aspiration level of α, if a DMU is efficient in one stage as a follower, then it is also efficient in that stage as a leader. Similarly, at any fixed aspiration level α, if a DMU is not efficient in one stage as a leader, then it is also not efficient in that stage as a follower. For instance, at the aspiration level $\alpha = 0.2$, DMU 1 is found to be stochastically efficient in the second stage as a follower. It is also stochastically efficient in the second stage as a leader. Similarly, at the aspiration level of $\alpha = 0.2$, since DMU 1 is not stochastically efficient as the leader in the first stage, it is also not stochastically efficient in the first stage as the follower.

Tables 5.5 and 5.6 present the stochastic efficiency scores of the DMUs in each stage as the leader or the follower, at the confidence level of $\alpha = 0.1$, for three different aspiration levels ($\beta = 1, 0.9, 0.8$). One can see that since DMU 15 is stochastically efficient in both the stages, it is efficient with the highest probability of $\beta = 1$. Moreover, since DMU 2 is satisficing inefficient at the lower aspiration

Table 5.1 Estimated mean values and standard deviations for 16 banks

DMU	Employees (10^3)	Fixed assets(RMB 10^8)	Expenses(RMB 10^8)	Deposits(RMB 10^8)	Interbank deposits(RMB 10^8)	Loan(RMB 10^8)	Profit(RMB 10^8)
1	(21.39,5.39)	(8.65,0.92)	(13.24,3.74)	(935.41,310.39)	(108.61,103.81)	(651.35,259.78)	(13.32,4.97)
2	(249.28,13.38)	(90.05,12.14)	(66.07,9.17)	(4925.35,1058.49)	(603.39,320.35)	(3189.65,1076.21)	(65.07,17.79)
3	(441.88,5.40)	(104.29,23.62)	(94.26,5.81)	(5989.71,1139.97)	(289.77,170.88)	(3014.98,451.41)	(51.45,29.15)
4	(49.29,0.57)	(9.91,0.06)	(8.54,1.07)	(183.29,53.73)	(119.83,56.46)	(1206.90,244.25)	(1.63,0.77)
5	(19.85,3.50)	(3.98,0.53)	(14.80,4.15)	(785.79,238.77)	(120.24,37.66)	(646.48,181.90)	(7.89,3.47)
6	(300.30,1.75)	(64.48,9.76)	(99.19,16.94)	(6375.92,1462.90)	(447.46,230.58)	(3683.58,821.06)	(92.64,26.62)
7	(16.99,2.54)	(8.18,2.66)	(8.30,1.86)	(523.90,115.67)	(121.69,78.56)	(450.52,128.52)	(7.32,2.28)
8	(385.61,17.29)	(80.30,12.33)	(91.51,15.73)	(8223.45,1558.40)	(592.61,236.65)	(4436.01,884.11)	(111.15,35.27)
9	(36.92,7.74)	(11.68,2.66)	(20.34,6.08)	(1250.65,367.00)	(115.79,67.66)	(852.75,269.32)	(20.95,5.99)
10	(19.54,5.89)	(3.39,0.30)	(10.35,2.84)	(558.34,80.84)	(182.91,54.44)	(489.99,161.84)	(11.39,4.12)
11	(10.38,1.63)	(1.68,0.19)	(5.22,1.32)	(254.97,50.46)	(36.06,24.25)	(281.71,78.97)	(0.61,1.95)
12	(17.60,4.39)	(6.26,0.70)	(12.68,2.74)	(805.30,269.91)	(222.44,98.13)	(681.27,147.24)	(12.52,4.96)
13	(77.73,8.60)	(22.74,2.07)	(23.56,2.65)	(1674.17,337.22)	(483.08,178.37)	(1298.78,386.33)	(28.52,8.23)
14	(11.11,1.86)	(3.64,0.20)	(7.29,1.62)	(394.35,72.21)	(87.01,29.12)	(345.67,70.81)	(3.07,1.02)
15	(6.22,1.05)	(12.44,7.93)	(5.50,2.05)	(242.99,108.07)	(115.45,78.89)	(2840.68,734.78)	(20.76,4.79)
16	(14.19,1.04)	(3.29,0.21)	(6.53,1.18)	(337.37,65.30)	(51.05,9.38)	(298.34,71.92)	(2.78,1.81)

Note. For (a,b), a is the mean value, and b is the standard deviation

Table 5.2 The efficiency scores obtained via the stochastic CCR model

	Confidence level				
	0.1	0.2	0.3	0.4	0.5
DMU	$\mathbf{e}^{*(\mathbf{o})}$	$\mathbf{e}^{*(\mathbf{o})}$	$\mathbf{e}^{*(\mathbf{o})}$	$\mathbf{e}^{*(\mathbf{o})}$	$\mathbf{e}^{*(\mathbf{o})}$
1	0.012	0	0.2773	0.0273	0.0018
2	0	0.0051	0.0743	0.1944	0.3656
3	0	0	0	0.0014	0.1264
4	0.0343	0.0895	0.1655	0.2645	0.376
5	0.0209	0.1127	0.1733	0.3373	0.4589
6	0	0	0.0037	0.1115	0.3841
7	0.0935	0.1799	0.269	0.3624	0.4494
8	0	0	0.0052	0.1147	0.3888
9	0	0	0.0299	0.2185	0.4446
10	0.038	0.1815	0.3	0.4	0.5
11	0.0562	0.1465	0.2241	0.346	0.4299
12	0.1	0.2	0.3	0.3679	0.4893
13	0.0887	0.102	0.2591	0.1498	0.4584
14	0.0013	0.0328	0.1133	0.2517	0.416
15	0.1	0.2	0.3	0.4	0.5
16	0	0	0	0.0042	0.2356

levels, it is stochastically inefficient with the highest probability of $\beta = 1$. This is because DMU 2 exhibits the lowest efficiency scores in both stages.

Figure 5.2 exhibits the extent of the difference in efficiency scores of a DMU (being considered the leader in Stage 2) with respect to the efficient DMUs, at different confidence levels. As one can see, when α increases, the difference in efficiency scores between inefficient and efficient units also increases. Similar results can be obtained when DMUs are assumed as followers in Stage 2. Figure 5.3 shows the extent of the difference in efficiency scores of a DMU (being considered the follower in Stage 2) with respect to the efficient DMUs, at different confidence levels.

The difference in efficiency scores in Stage 1 as the leader for each DMU with respect to the efficient DMU at $\alpha = 0.1$, under three aspiration levels β ($\beta = 1, 0.9$, and 0.8, as indicated in Fig. 5.4). As it can be seen, with a decrease in the aspiration level, the difference in efficiency scores between the efficient and inefficient units decreases.

5.3 The Cooperative Models

In real-world situations, there are many cases in which there is a need for perfect coordination between the stages of a production system to achieve maximal overall performance (Maghbouli et al., 2014). For example, marketing and production departments need to work together to maximize the company's profit (Maghbouli

Table 5.3 The efficiency scores of the stochastic network DEA model (Stage 1 is the leader, Stage 2 is the follower) ($\beta = 1$)

	Confidence level (α)									
	0.1		0.2		0.3		0.4		0.5	
DMU	$e_{1l}^{*(o)}$	$e_{2f}^{*(o)}$	$e_{1l}^{*(o)}$	$e_{2f}^{*(o)}$	$e_{1l}^{*(o)}$	$e_{2f}^{*(o)}$	$e_{1l}^{*(o)}$	$e_{2f}^{*(o)}$	$e_{1l}^{*(o)}$	$e_{2f}^{*(o)}$
1	0.1	0.001989	0.2	0.002498	0.3	0.002904	0.4	0.003265	0.5	0.003599
2	0.010079	1.253e-5	0.027965	1.62e-5	0.054215	1.995e-5	0.090625	2.42e-5	0.094625	2.936e-5
3	0.000013481	8.825e-8	0.000020895	8.884e-7	0.00003843	1.269e-6	0.000079629	9.8e-7	0.14003	1.227e-6
4	0.00047558	8.825e-8	0.003031	0.03402	0.009976	0.04464	0.024617	0.05446	0.00018314	0.06412
5	0.1	0.0005837	0.2	0.0006019	0.3	0.0006101	0.4	0.000613	0.054529	0.0009529
6	1.0796e-7	6.674e-6	1.1407e-6	6.705e-6	8.6513e-6	6.723e-6	0.00020629	6.736e-6	0.5	0.5
7	0.010557	6.674e-6	0.033537	0.07758	0.068877	0.09224	0.12006	0.01568	0.0024272	0.06091
8	0.1	3.417e-7	0.2	4.649e-7	0.3	6.064e-7	0.4	7.89e-7	0.20433	1.047e-6
9	0.00063077	0.000342	0.00063511	0.0003467	0.00063874	0.0003497	0.00064224	0.0003519	0.00064593	0.0003539
10	0.1	0.0001692	0.2	0.0003899	0.3	0.0003899	0.4	0.0003899	0.5	0.0003899
11	0.00077071	2.176e-7	0.00079066	2.176e-7	0.00080738	2.176e-7	0.0015434	0.0002038	0.0058468	0.003821
12	0.1	0.01005	0.2	0.01441	0.3	0.01559	0.4	0.0165	0.5	0.01728
13	0.1	0.001946	0.2	0.004431	0.3	0.005641	0.4	0.006071	0.5	0.006383
14	0.00034526	2.365e-8	0.00043653	2.365e-8	0.00053077	2.365e-8	0.0006431	2.365e-8	0.00096479	0.001324
15	0.1	0.1	0.2	0.2	0.3	0.3	0.4	0.4	0.5	0.5
16	1.566e-8	0.1	1.566e-8	0.2	1.566e-8	0.3	1.566e-8	0.4	1.566e-8	0.5

Table 5.4 The efficiency scores of the stochastic network DEA model (Stage 2 is the leader, Stage 1 is the follower) ($\beta = 1$)

DMU	Confidence level (α)									
	0.1		0.2		0.3		0.4		0.5	
	$e_{2l}^{*(o)}$	$e_{1f}^{*(o)}$	$e_{2l}^{*(o)}$	$e_{1f}^{*(o)}$	$e_{2l}^{*(o)}$	$e_{1f}^{*(o)}$	$e_{2l}^{*(o)}$	$e_{1f}^{*(o)}$	$e_{2l}^{*(o)}$	$e_{1f}^{*(o)}$
1	0.1	0.0002	0.2	0.0002	0.25717	0.0002	0.27749	0.0002	0.29236	0.0002
2	0.012382	2.902e-13	0.052558	2.902e-13	0.085788	2.902e-13	0.099254	8.76e-13	0.29236	0.000891
3	0.0058344	5.092e-6	0.082695	5.055e-6	0.10308	5.054e-6	0.12047	5.042e-6	0.10964	2.578e-8
4	0.019739	2.065e-5	0.034017	0.003031	0.044637	0.009976	0.054463	0.02462	0.16877	5.045e-6
5	0.00067829	0.0001811	0.00081729	0.0001811	0.00087061	0.0001811	0.00091402	0.0001811	0.064123	0.05453
6	0.0065523	2.38e-9	0.084436	2.38e-9	0.28007	2.38e-9	0.4	2.38e-9	0.00095288	0.0001811
7	0.041692	0.008423	0.077578	0.03354	0.092237	0.06888	0.098233	0.1181	0.10281	2.38e-9
8	0.00083694	2.986e-9	0.016903	2.986e-9	0.065182	2.986e-9	0.12863	2.986e-9	0.21407	0.1851
9	0.022329	0.0004112	0.11261	0.0004108	0.23908	0.0004109	0.31289	0.0004109	0.37504	2.986e-9
10	0.00016917	0.1	0.00038993	0.2	0.00038993	0.3	0.00038993	0.4	0.00038993	0.0004108
11	0.0377	3.857e-5	0.092438	3.834e-5	0.10377	3.846e-5	0.11293	3.845e-5	0.12106	5.201e-5
12	0.010052	0.1	0.01441	0.2	0.015586	0.3	0.016496	0.4	0.017278	0.5
13	0.001946	0.1	0.0044309	0.2	0.0056414	0.3	0.0060708	0.4	0.0063828	0.5
14	0.00061566	3.452e-6	0.0019848	3.448e-6	0.0022695	3.44e-6	0.0025041	3.351e-5	0.0027153	0.0003028
15	0.1	0.1	0.2	0.2	0.3	0.3	0.4	0.4	0.5	0.5
16	1.1928e-7	1.566e-8	1.1928e-7	1.566e-8	1.1928e-7	1.566e-8	1.1928e-7	1.566e-8	1.1928e-7	1.566e-8

Table 5.5 The efficiency scores of the stochastic network DEA model (Stage 1 is the leader, Stage 2 is the follower) ($\alpha = 0.1$)

	Aspiration level (β)					
	1		0.9		0.8	
DMU	$e_{1l}^{*(o)}$	$e_{2f}^{*(o)}$	$e_{1l}^{*(o)}$	$e_{2f}^{*(o)}$	$e_{1l}^{*(o)}$	$e_{2f}^{*(o)}$
1	0.1	0.001989	0.1	0.001989	0.1	0.001989
2	0.010079	1.25E-05	0.010079	1.25E-05	0.010079	1.25E-05
3	1.35E-05	8.82E-08	1.35E-05	8.82E-08	1.35E-05	8.82E-08
4	0.000476	8.82E-08	0.000476	8.82E-08	0.000476	8.82E-08
5	0.1	0.000584	0.1	0.000584	0.1	0.000584
6	1.08E-07	6.67E-06	1.08E-07	6.67E-06	1.08E-07	6.67E-06
7	0.010557	6.67E-06	0.010557	6.67E-06	0.010557	6.67E-06
8	0.1	3.42E-07	0.1	3.42E-07	0.1	3.42E-07
9	0.000631	0.000342	0.000631	0.000342	0.000631	0.000342
10	0.1	0.000169	0.1	0.000169	0.1	0.000169
11	0.000771	2.18E-07	0.000771	2.18E-07	0.000771	2.18E-07
12	0.1	0.010052	0.1	0.010052	0.1	0.010052
13	0.1	0.001946	0.1	0.001946	0.1	0.001946
14	0.000345	2.37E-08	0.000345	2.37E-08	0.000345	2.37E-08
15	0.1	0.1	0.1	0.1	0.1	0.1
16	1.57E-08	1.00E-01	1.57E-08	1.00E-01	1.57E-08	1.00E-01

Table 5.6 The efficiency scores of the stochastic network DEA model (Stage 2 is the leader, Stage 1 is the follower) ($\alpha = 0.1$)

	Aspiration level (β)					
	1		0.9		0.8	
DMU	$e_{2l}^{*(o)}$	$e_{1f}^{*(o)}$	$e_{2l}^{*(o)}$	$e_{1f}^{*(o)}$	$e_{2l}^{*(o)}$	$e_{1f}^{*(o)}$
1	0.1	0.0002	0.1	0.0002	0.1	0.0002
2	0.012382	2.90E-13	0.012382	2.90E-13	0.012382	2.90E-13
3	5.83E-03	5.09E-06	5.83E-03	5.10E-06	5.83E-03	5.05E-06
4	0.019739	2.07E-05	0.019739	2.07E-05	0.019739	2.07E-05
5	0.000678	0.000181	0.000678	0.000181	0.000678	0.000181
6	6.55E-03	2.38E-09	6.55E-03	2.38E-09	6.55E-03	2.38E-09
7	0.041692	8.42E-03	0.041692	8.42E-03	0.041692	8.42E-03
8	0.000837	2.99E-09	0.000837	2.99E-09	0.000837	2.99E-09
9	0.022329	0.000411	0.022329	0.000411	0.022329	0.000411
10	0.000169	0.1	0.000169	0.1	0.000169	0.1
11	0.0377	3.86E-05	0.0377	3.86E-05	0.0377	3.84E-05
12	0.010052	0.1	0.010052	0.1	0.010052	0.1
13	0.001946	0.1	0.001946	0.1	0.001946	0.1
14	0.000616	3.45E-06	0.000616	3.46E-06	0.000616	3.47E-06
15	0.1	0.1	0.1	0.1	0.1	0.1
16	1.19E-07	1.57E-08	1.19E-07	1.57E-08	1.19E-07	1.57E-08

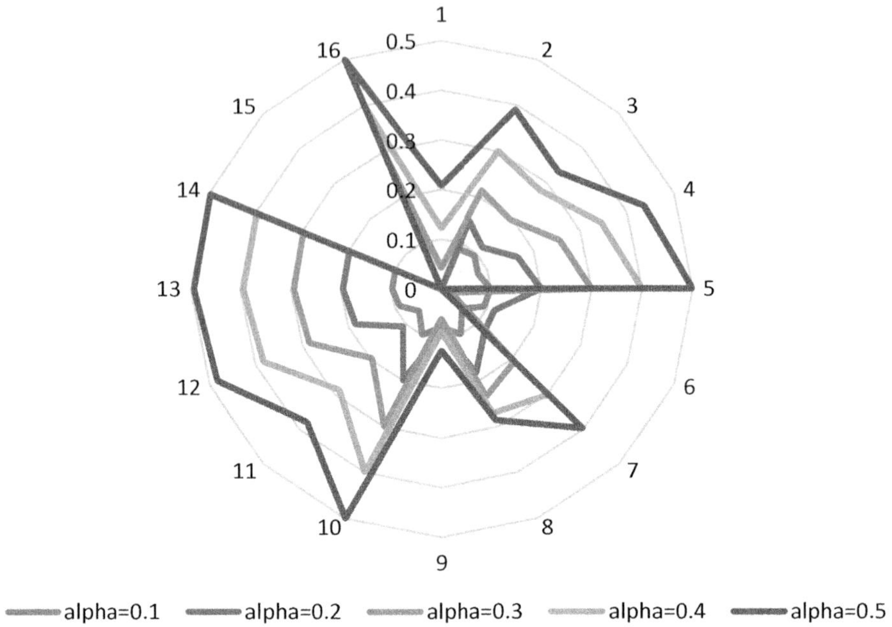

Fig. 5.2 The difference in efficiency scores in Stage 2 as the leader for the efficient DMUs, at different confidence levels

et al., 2014). In this section, we first review the deterministic centralized two-stage DEA model and then provide its dual formulation.

5.3.1 Deterministic Centralized Model

Consider Fig. 5.1 and suppose that there are n independent DMUs, denoted by DMU_j ($j = 1, 2, \ldots, n$), in which the first stage consumes the input vector $x_j = (x_{1j}, x_{2j}, \ldots, x_{Mj})^T$ to produce the intermediate output vector $z_j = (z_{1j}, z_{2j}, \ldots, z_{Dj})^T$, which is used as input in the second stage to produce the final output vector $y_j = (y_{1j}, y_{2j}, \ldots, y_{Sj})^T$. In the following, we present a centralized model based on the concept of cooperative game theory. In our approach, the two-stage process is viewed as one stage where the two stages jointly determine one optimal plan to maximize the total efficiency of the whole system (Maghbouli et al., 2014). To measure the relative efficiency of DMU_o, the following multiplier model is proposed:

$$\text{Max } \frac{Uy_o}{Vx_o} \tag{5.37}$$

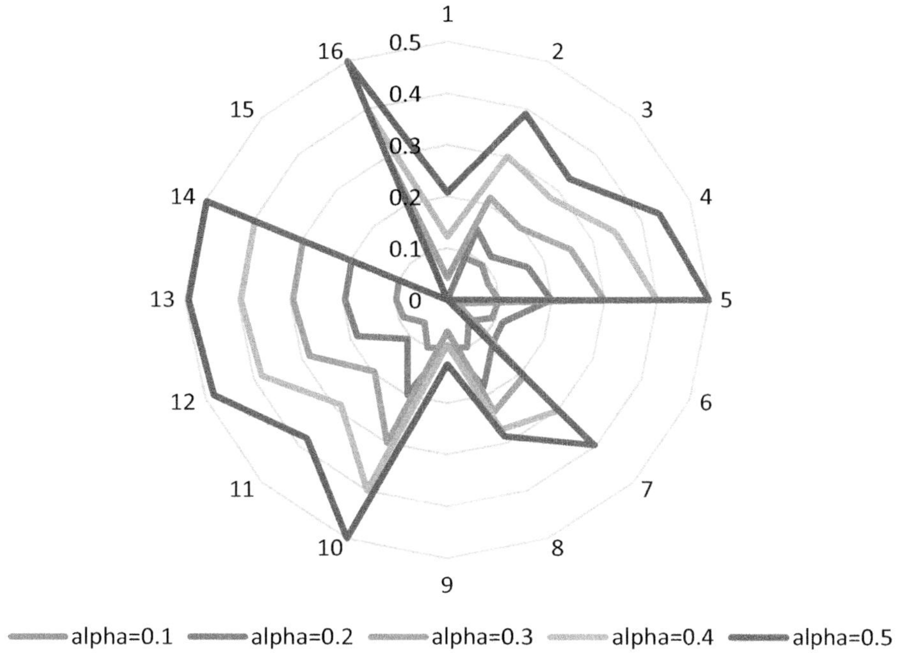

Fig. 5.3 The difference in efficiency scores in Stage 2 as the follower for the efficient DMUs, at different confidence levels

Fig. 5.4 The difference in efficiency scores in Stage 1 as the leader for the efficient DMUs, under different aspiration levels

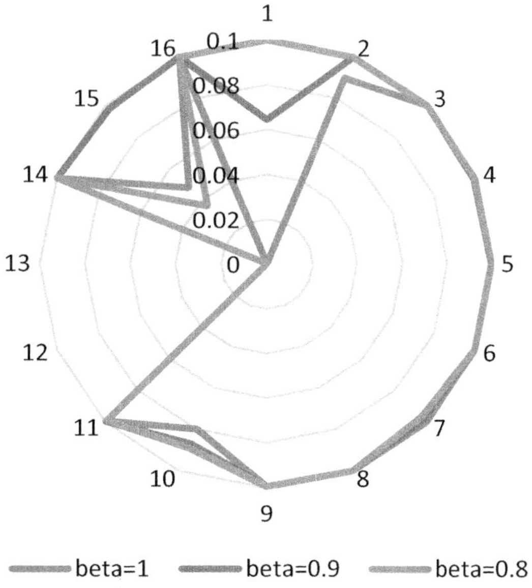

$$\text{s.t.} \quad \frac{Uy_j}{W\tilde{z}_j} \leq 1 \ (\forall j),$$
$$\frac{Wz_j}{Vx_j} \leq 1 \ (\forall j),$$
$$\frac{Uy_j}{Vx_j} \leq 1 \ (\forall j),$$
$$(W, U, V) \geq 0,$$

where *V*, *W*, and *U* are the weight vectors of the input, intermediate measure, and output, respectively. In the objective function of (5.37), the ratio of the weighted sum of final outputs to the weighted sum of initial inputs is maximized. The constraints guarantee that the ratio of the weighted sum of the outputs to the weighted sum of the inputs in both stages and the whole system does not exceed unity. Model (5.37) is a linear fractional programming problem. By using Charnes and Cooper's (1962) transformation, model (5.37) can be transformed into the following linear form:

$$\text{Max } Uy_o \tag{5.38}$$
$$\text{s.t.} \quad Vx_o = 1,$$
$$Uy_j - Wz_j \leq 0 \ (\forall j),$$
$$Wz_j - Vx_j \leq 0 \ (\forall j),$$
$$Uy_j - Vx_j \leq 0 \ (\forall j),$$
$$W, U, V \geq 0.$$

The dual formulation of model (5.38) can be set up as

$$\text{Min } \theta \tag{5.39}$$
$$\text{s.t.} \quad \sum_{j=1}^{n} (\lambda_j + \gamma_j) x_{ij} \leq \theta x_{io} \ (\forall i),$$
$$\sum_{j=1}^{n} (\lambda_j + \xi_j) y_{rj} \geq y_{ro} \ (\forall r),$$
$$\sum_{j=1}^{n} (\gamma_j - \xi_j) z_{dj} \geq 0 \ (\forall d),$$
$$\lambda_j, \gamma_j, \xi_j \geq 0, \quad (\forall j).$$

In model (5.39), $(\lambda_j + \gamma_j)$, $(\gamma_j - \xi_j)$, and $(\lambda_j + \xi_j)$ are the multipliers related to the inputs, intermediate measures, and final outputs, respectively.

5.3.2 Stochastic Centralized Network Model

In this section, we develop a stochastic centralized DEA model for a decision process with a two-stage structure. Let us assume that all initial inputs, intermediate products, and final outputs be jointly normally distributed. Then, the stochastic version of model (5.39) in a chance-constrained form can be formulated as

$$
\begin{aligned}
e_{SCN}^{(\alpha)} &= \text{Min } \theta \qquad (5.40)\\
\text{s.t.} \quad & \mathbb{P}\left(\sum_{j=1}^{n}(\lambda_j+\gamma_j)\widetilde{x}_{ij} \le \theta\widetilde{x}_{io}\right) \ge 1-\alpha \ (\forall i),\\
& \mathbb{P}\left(\sum_{j=1}^{n}(\lambda_j+\xi_j)\widetilde{y}_{rj} \ge \widetilde{y}_{ro}\right) \ge 1-\alpha \ (\forall r),\\
& \mathbb{P}\left(\sum_{j=1}^{n}(\gamma_j-\xi_j)\widetilde{z}_{dj} \ge 0\right) \ge 1-\alpha \ (\forall d),\\
& \lambda_j, \gamma_j, \xi_j \ge 0 \ (\forall j).
\end{aligned}
$$

In model (5.40), $\widetilde{x}_j = \left(\widetilde{x}_{1j}, \widetilde{x}_{2j}, \ldots, \widetilde{x}_{Mj}\right)^T$, $\widetilde{z}_j = \left(\widetilde{z}_{1j}, \widetilde{z}_{2j}, \ldots, \widetilde{z}_{Dj}\right)^T$, and $\widetilde{y}_j = \left(\widetilde{y}_{1j}, \widetilde{y}_{2j}, \ldots, \widetilde{y}_{Sj}\right)^T$ are the random initial input, intermediate product, and final output vectors, respectively, and $x_j = (x_{1j}, x_{2j}, \ldots, x_{Mj})^T$, $z_j = (z_{1j}, z_{2j}, \ldots, z_{Dj})^T$, and $\mathrm{y}_j = (\mathrm{y}_{1j}, \mathrm{y}_{2j}, \ldots, \mathrm{y}_{Sj})^T$ are their corresponding mean vectors. And, $\mathbb{P}$ means probability and $\alpha \in [0, 1]$ is a pre-specified scalar that represents the allowable chance (risk) of failing to satisfy the constraints with which it is associated.

Definition 5.1. For a predetermined level α, DMU_o is stochastically efficient if and only if $e_{SCN}^{(\alpha)} = 1$.

Suppose that T_{stage1}^s and T_{stage2}^s are, respectively, the stochastic technology (production possibility) sets of the first and second stages, which are defined as

$$
T_{stage1}^s = \left\{ \left(\widetilde{x}^T, \widetilde{z}^T\right)^T \in \mathbb{R}_+^{M+D} \,|\, \exists \left(\widetilde{x}_j^T, \widetilde{z}_j^T\right)^T \in D_j(1-\alpha_j) \ \text{and} \ \lambda_j, \gamma_j \ge 0 \ (\forall j) \right.
$$
$$
\left. \text{such that} \sum_{j=1}^{n}(\lambda_j+\gamma_j)\widetilde{x}_j \le \widetilde{x} \ \text{and} \ \sum_{j=1}^{n}(\lambda_j+\gamma_j)\widetilde{z}_j \ge \widetilde{z} \right\},
$$

$$T^{s}_{stage2} = \left\{ \left(\widetilde{z}^{T}, \widetilde{y}^{T}\right)^{T} \in \mathbb{R}^{D+S}_{+} \mid \exists \left(\widetilde{z}^{T}_{j}, \widetilde{y}^{T}_{j}\right)^{T} \in D'_{j}(1-\alpha_{j}) \text{ and } \lambda_{j}, \xi_{j} \geq 0 \ (\forall j) \right.$$
$$\left. \text{such that } \sum_{j=1}^{n} (\lambda_{j} + \xi_{j}) \widetilde{z}_{j} \leq \widetilde{z} \text{ and } \sum_{j=1}^{n} (\lambda_{j} + \xi_{j}) \widetilde{y}_{j} \geq \widetilde{y}, \right\},$$

where $D_j(1 - \alpha_j)$ and $D_j'(1 - \alpha_j)$ represent the confidence regions for Stages 1 and 2 for DMU_j, respectively. These confidence regions are defined by Olesen and Petersen (2000) as

$$D_{j}(1-\alpha_{j}) = \left\{ \left(\widetilde{x}^{T}, \widetilde{z}^{T}\right)^{T} \in \mathbb{R}^{M+D}_{+} \mid \left[(\widetilde{x} - x_{j})^{T}, (\widetilde{z} - z_{j})^{T} \right] \right.$$
$$\left. \Lambda_{j}^{-1} \left[(\widetilde{x} - x_{j})^{T}, (\widetilde{z} - z_{j})^{T} \right]^{T} \leq c_{j}^{2} \right\},$$
$$D_{j}'(1-\alpha_{j}) = \left\{ \left(\widetilde{z}^{T}, \widetilde{y}^{T}\right)^{T} \in \mathbb{R}^{D+S}_{+} \mid \left[(\widetilde{z} - z_{j})^{T}, (\widetilde{y} - y_{j})^{T} \right] \right.$$
$$\left. \Lambda_{j}'^{-1} \left[(\widetilde{z} - z_{j})^{T}, (\widetilde{y} - y_{j})^{T} \right]^{T} \leq c_{j}^{2} \right\},$$

where Λ_j^{-1} and $\Lambda_j'^{-1}$ are the inverses of the variance–covariance matrix of $(\widetilde{x}_j, \widetilde{z}_j)$ and $(\widetilde{z}_j, \widetilde{y}_j)$, respectively. Moreover, c_j is determined by $\mathbb{P}\left(\chi^2_{M+D} \leq c_j^2\right) = 1 - \alpha_j$, and χ^2_{M+D} is the Chi-square random variable with $M + D$ degrees of freedom.

Olesen and Petersen (2000) clarified that the random realizations of DMU_j fall within the confidence region $D_j(1 - \alpha_j)$ positions inside the technology set if $\alpha \leq 0.5$ (Wei et al., 2014). So, T^s_{stage1} and T^s_{stage2} represent the envelopment of n confidence regions $D_j(1 - \alpha_j)$ and $D_j'(1 - \alpha_j)$ for $j = 1, \ldots, n$, respectively (Olesen & Petersen, 2000). Then, the stochastic technology set of the stochastic centralized network DEA model (5.40) $\left(T^s_{SCN}\right)$ can be represented as

$$T^{s}_{SCN} = \left\{ (\widetilde{x}, \widetilde{y}) \mid \exists \widetilde{z} \in D_{j}(1-\alpha_{j}) \cap D_{j}'(1-\alpha_{j}) : (\widetilde{x}, \widetilde{z}) \in T^{s}_{stage1}, (\widetilde{z}, \widetilde{y}) \in T^{s}_{stage2} \right\}$$
$$= \left\{ \begin{array}{l} (\widetilde{x}, \widetilde{y}) \mid \exists \widetilde{z} \in D_{j}(1-\alpha_{j}) \cap D'_{j}(1-\alpha_{j}) : \sum_{j=1}^{n} (\lambda_{j} + \gamma_{j}) \widetilde{x}_{j} \leq \widetilde{x}, \\ \sum_{j=1}^{n} \gamma_{j} \widetilde{z}_{j} \geq \sum_{j=1}^{n} \xi_{j} \widetilde{z}_{j}, \sum_{j=1}^{n} (\lambda_{j} + \xi_{j}) \widetilde{y}_{j} \geq \widetilde{y}, \lambda_{j}, \gamma_{j}, \xi_{j} \geq 0 \end{array} \right\}.$$

As it can be seen, T^s_{SCN} is equivalent to the stochastic technology set under the centralized control organization mechanism proposed by Zhou et al. (2017). Model (5.40) calculates the stochastic efficiency of the whole system in the spirit of Farrell's (1957) input radial measure.

5.3.3 Transformation to Deterministic Equivalent Linear Models

Following Cooper et al. (1996), to evaluate the relative efficiency of DMU_o, one can transform the chance-constrained model (5.40) into a deterministic form. To this end, consider the ith input constraint of model (5.40):

$$\mathbb{P}\left(\sum_{j=1}^{n}(\lambda_j+\gamma_j)\widetilde{x}_{ij}-\theta\widetilde{x}_{io}\leq 0\right)\geq 1-\alpha. \tag{5.41}$$

By introducing the slack variable $\varepsilon_i \geq 0$, the inequality constraint (5.41) is transformed into the following equality form:

$$\mathbb{P}\left(\sum_{j=1}^{n}(\lambda_j+\gamma_j)\widetilde{x}_{ij}-\theta\widetilde{x}_{io}\leq 0\right)=1-\alpha+\varepsilon_i. \tag{5.42}$$

Note 5.1. Suppose X is a random variable and a, b, and c are constants. If $a \geq b$ and $P(X \leq a) = c$, then $P(X \leq b) = d$, where $d \leq c$.

Taking Note 5.1 into consideration, and introducing the ith input slack $s_i^- \geq 0$ in (5.52), we have

$$\mathbb{P}\left(\sum_{j=1}^{n}(\lambda_j+\gamma_j)\widetilde{x}_{ij}-\theta\widetilde{x}_{io}\leq -s_i^-\right)=1-\alpha. \tag{5.43}$$

Define $\widetilde{h}_i=\sum_{j=1}^{n}(\lambda_j+\gamma_j)\widetilde{x}_{ij}-\theta\widetilde{x}_{io}$. Then, $\widetilde{h}_i \sim N\left(h_i,\left(\sigma_i^I(\lambda,\gamma,\theta)\right)^2\right)$, where $h_i=E\left(\widetilde{h}_i\right)=E\left(\sum_{j=1}^{n}(\lambda_j+\gamma_j)\widetilde{x}_{ij}-\theta\widetilde{x}_{io}\right)=\sum_{j=1}^{n}(\lambda_j+\gamma_j)x_{ij}-\theta x_{io}$ and

$$\left(\sigma_i^I(\lambda,\gamma,\theta)\right)^2=Var\left(\widetilde{h}_i\right)=Var\left(\sum_{j=1}^{n}(\lambda_j+\gamma_j)\widetilde{x}_{ij}-\theta\widetilde{x}_{io}\right)$$

$$=Var\left(\sum_{j=1}^{n}(\lambda_j+\gamma_j)\widetilde{x}_{ij}\right)+var(\theta\widetilde{x}_{io})-2Cov\left(\sum_{j=1}^{n}(\lambda_j+\gamma_j)\widetilde{x}_{ij},\theta\widetilde{x}_{io}\right)$$

$$=\sum_{j=1}^{n}\sum_{k=1}^{n}(\lambda_j+\gamma_j)(\lambda_k+\gamma_k)Cov(\widetilde{x}_{ij},\widetilde{x}_{ik})+\theta^2 Var(\widetilde{x}_{io})-2\theta\sum_{j=1}^{n}$$
$$\times(\lambda_j+\gamma_j)cov(\widetilde{x}_{ij},\widetilde{x}_{io}).$$

Now, by considering (5.43), we have $\mathbb{P}\left(\widetilde{h}_i \le -s_i^-\right) = 1-\alpha \Rightarrow \mathbb{P}\left(\frac{\widetilde{h}_i - h_i}{\sigma_i^I(\lambda,\gamma,\theta)} \le \frac{-s_i^- - h_i}{\sigma_i^I(\lambda,\gamma,\theta)}\right) = 1-\alpha$, where $\widetilde{Z}_i = \frac{\widetilde{h}_i - h_i}{\sigma_i^I(\lambda,\gamma,\theta)} \sim N(0,1)$. Therefore, $\mathbb{P}\left(\widetilde{Z}_i \le \frac{-s_i^- - h_i}{\sigma_i^I(\lambda,\gamma,\theta)}\right) = 1-\alpha \Rightarrow \mathbb{P}\left(\widetilde{Z}_i \le \frac{s_i^- + h_i}{\sigma_i^I(\lambda,\gamma,\theta)}\right) = \alpha \Rightarrow \Phi\left(\frac{s_i^- + h_i}{\sigma_i^I(\lambda,\gamma,\theta)}\right) = \alpha$, where $\Phi(\cdot)$ is the standard normal distribution function. Eventually, according to the invertibility of $\Phi(.)$, we have

$$\frac{s_i^- + h_i}{\sigma_i^I(\lambda,\gamma,\theta)} = \Phi^{-1}(\alpha) \Rightarrow h_i + s_i^- - \Phi^{-1}(\alpha)\sigma_i^I(\lambda,\theta) = 0$$

$$\Rightarrow \sum_{j=1}^{n}(\lambda_j+\gamma_j)x_{ij} + s_i^- - \Phi^{-1}(\alpha)\sigma_i^I(\lambda,\gamma,\theta) = \theta x_{io}.$$

In a similar way, the rth output constraint in (5.40) can be transformed into the following form: $\sum_{j=1}^{n}(\lambda_j+\xi_j)y_{rj} - s_i^+ + \Phi^{-1}(\alpha)\sigma_r^O(\lambda,\xi) = y_{ro}$, where $\left(\sigma_r^O(\lambda,\xi)\right)^2 = Var\left(\widetilde{h}_i\right) = \sum_{j=1}^{n}\sum_{k=1}^{n}(\lambda_j+\xi_j)(\lambda_k+\xi_k)Cov\left(\widetilde{y}_{rj},\widetilde{y}_{rk}\right) + Var(\widetilde{y}_{ro}) - 2\sum_{j=1}^{n}(\lambda_j+\xi_j)cov\left(\widetilde{y}_{rj},\widetilde{y}_{ro}\right)$.

Moreover, for the third constraint, we have $\sum_{j=1}^{n}(\xi_j-\gamma_j)z_{dj} + s_d^- - \Phi^{-1}(\alpha)\sigma_d^z(\xi,\gamma) = 0$, where $\left(\sigma_d^z(\xi,\gamma)\right)^2 = Var\left(\sum_{j=1}^{n}(\xi_j-\gamma_j)\widetilde{z}_{dj}\right) = \sum_{j=1}^{n}\sum_{k=1}^{n}(\xi_j-\gamma_j)(\xi_k-\gamma_k)cov\left(\widetilde{z}_{dj},\widetilde{z}_{dk}\right)$.

Consequently, model (5.40) can be transformed into the following quadratic program:

$$e_{SCN}^{(\alpha)} = \text{Min }\theta \tag{5.44}$$

$$\text{s.t.} \quad \sum_{j=1}^{n}(\lambda_j+\gamma_j)x_{ij} + s_i^- - \Phi^{-1}(\alpha)v_i = \theta x_{io} \ (\forall i),$$

$$\sum_{j=1}^{n}(\lambda_j+\xi_j)y_{rj} - s_r^+ + \Phi^{-1}(\alpha)u_r = y_{ro} \ (\forall r),$$

$$\sum_{j=1}^{n}(\gamma_j-\xi_j)z_{dj} + s_d^- - \Phi^{-1}(\alpha)w_d = 0 \ (\forall d),$$

$$v_i^2 = \sum_{j=1}^{n}\sum_{k=1}^{n}(\lambda_j+\gamma_j)(\lambda_k+\gamma_k)cov\left(\widetilde{x}_{ij},\widetilde{x}_{ik}\right) + \theta^2 var(\widetilde{x}_{io}) - 2\theta\sum_{j=1}^{n} \times (\lambda_j+\gamma_j)cov\left(\widetilde{x}_{ij},\widetilde{x}_{io}\right) \ (\forall i),$$

$$u_r^2 = \sum_{j=1}^{n}\sum_{k=1}^{n}(\lambda_j+\xi_j)(\lambda_k+\xi_k)cov(\widetilde{y}_{rj},\widetilde{y}_{rk}) + var(\widetilde{y}_{ro}) - 2\sum_{j=1}^{n}$$
$$\times(\lambda_j+\xi_j)cov(\widetilde{y}_{rj},\widetilde{y}_{ro})\ (\forall r),$$

$$w_d^2 = \sum_{j=1}^{n}\sum_{k=1}^{n}(\xi_j-\gamma_j)(\xi_k-\gamma_k)cov(\widetilde{z}_{dj},\widetilde{z}_{dk})\ (\forall d),$$

$$\gamma,\lambda,\xi \geq 0.$$

Theorem 5.9 Model (5.44) is always feasible at all the confidence levels of α.

Proof It is easy to show that $\theta = 1, \lambda_o + \gamma_o = 1,\ \lambda_o + \xi_o = 1, \lambda_j + \gamma_j = 0,\ \lambda_j + \xi_j = 0$, $\gamma_j - \xi_j = 0$ for all $j \neq o$ and that $v = w = u = 0$ is a feasible solution to model (5.44). ∎

Proposition 5.1 For $\alpha \leq 0.5$, $0 \leq e_{SCN}^{(\alpha)} \leq 1$.

Proof Given the feasible solution in Theorem 5.9, and that model (5.44) is a minimization program, we have $e_{SCN}^{(\alpha)} \leq 1$. Moreover, since $\alpha \leq 0.5$, then $-\Phi^{-1}(\alpha) \geq 0$ and $v \geq 0$. Taking the input constraints of model (5.44) into consideration, we have $\sum_{j=1}^{n}(\lambda_j+\gamma_j)x_{ij} \leq \theta x_{io} \Rightarrow 0 < \frac{\sum_{j=1}^{n}(\lambda_j+\gamma_j)x_{ij}}{x_{io}} \leq \theta$. Hence, $e_{SCN}^{(\alpha)} > 0$. ∎

Theorem 5.10 If $\alpha = 0.5$, then the objective function values of models (5.44) and (5.39) are the same.

Proof Since $\Phi^{-1}(\alpha) = 0$, the values of the objective function in both the models are the same. The proof is hence complete.

5.3.4 *Error Structure*

In this subsection, we employ the linearization approach by Cooper et al. (1998) regarding the error structure of data to transform the chance-constrained problem into an equivalent linear deterministic program. Toward this end, we suppose the inputs, intermediate measures, and final outputs of the DMUs are designated by single factors.

Let the inputs, intermediate products, and final outputs of the jth *DMU* be represented as $\widetilde{x}_{ij} = x_{ij} + a_{ij}\varepsilon_{ij}\ (\forall i)$, $\widetilde{z}_{dj} = z_{dj} + c_{dj}\tau_{dj}\ (\forall d)$, and $\widetilde{y}_{rj} = y_{rj} + b_{rj}\eta_{rj}\ (\forall r)$, respectively, where a_{ij}, b_{rj}, and c_{dj} are non-negative values. Further, let $\varepsilon_{ij} \sim N(0,\overline{\sigma}^2)$, $\eta_{rj} \sim N(0,\overline{\sigma}^2)$, and $\tau_{dj} \sim N(0,\overline{\sigma}^2)$. Then, $\widetilde{x}_{ij} \sim N\left(x_{ij},\overline{\sigma}^2a_{ij}^2\right)$, $\widetilde{y}_{rj} \sim N\left(y_{rj},\overline{\sigma}^2b_{ij}^2\right)$, and $\widetilde{z}_{dj} \sim N\left(z_{dj},\overline{\sigma}^2c_{dj}^2\right)$.

Now, suppose that for each DMU_j ($j = 1, \ldots, n$), we have $\tilde{\varepsilon}_i = \tilde{\varepsilon}_{ij}$, $\tilde{\eta}_r = \tilde{\eta}_{rj}$, and $\tilde{\tau}_d = \tilde{\tau}_{dj}$. Using the above assumptions, model (5.40) can then be transformed into the following form:

$$e_{SCN}^{(\alpha)} = \text{Min } \theta \tag{5.45}$$

$$\text{s.t.} \quad \sum_{j=1}^{n} (\lambda_j + \gamma_j) x_{ij} - \Phi^{-1}(\alpha)\, \overline{\sigma} \left| \sum_{j=1}^{n} (\lambda_j + \gamma_j) a_{ij} - \theta a_{io} \right| \le \theta x_{io} \ (\forall i),$$

$$\sum_{j=1}^{n} (\lambda_j + \xi_j) y_{rj} + \Phi^{-1}(\alpha)\, \overline{\sigma} \left| \sum_{j=1}^{n} (\lambda_j + \xi_j) b_{rj} - b_{ro} \right| \ge y_{ro} \ (\forall r),$$

$$\sum_{j=1}^{n} (\xi_j - \gamma_j) z_{dj} - \Phi^{-1}(\alpha)\, \overline{\sigma} \left| \sum_{j=1}^{n} (\xi_j - \gamma_j) c_{dj} \right| \le 0 \ (\forall d),$$

$$\lambda_j, \gamma_j, \xi_j \ge 0 \ (\forall j).$$

Due to the existence of the absolute values in the constraints, model (5.45) is non-linear. We will, therefore, be using the goal programming theory by Charnes and Cooper (1961, 1977) to convert this model into a linear form. To do this, we use the following transformations:

$$\left| \sum_{j=1}^{n} (\lambda_j + \gamma_j) a_{ij} - \theta a_{io} \right| = p_i^+ + p_i^- \ (\forall i),$$

$$\sum_{j=1}^{n} (\lambda_j + \gamma_j) a_{ij} - \theta a_{io} = p_i^+ - p_i^- \ (\forall i),$$

$$p_i^+ p_i^- = 0 \ (\forall i)$$

$$\left| \sum_{j=1}^{n} (\lambda_j + \xi_j) b_{rj} - b_{ro} \right| = q_r^+ + q_r^- \ (\forall r),$$

$$\sum_{j=1}^{n} (\lambda_j + \xi_j) b_{rj} - b_{ro} = q_r^+ - q_r^- \ (\forall r),$$

$$q_r^+ q_r^- = 0 \ (\forall r),$$

$$\left| \sum_{j=1}^{n} (\xi_j - \gamma_j) c_{dj} \right| = \delta_d^+ + \delta_d^- \ (\forall d),$$

$$\sum_{j=1}^{n} (\xi_j - \gamma_j) c_{dj} = \delta_d^+ - \delta_d^- \ (\forall d),$$

$$\delta_d^+ \delta_d^- = 0 \ (\forall i).$$

Using the above transformations, model (5.45) can be converted into the following form:

$$e_{SCN}^{(\alpha)} = \text{Min}\theta \tag{5.46}$$

$$\text{s.t.} \ \sum_{j=1}^{n} (\lambda_j + \gamma_j) x_{ij} - \Phi^{-1}(\alpha)\, \overline{\sigma}(p_i^+ + p_i^-) \le \theta x_{io} \ (\forall i),$$

$$\sum_{j=1}^{n} (\lambda_j + \gamma_j) a_{ij} - \theta a_{io} = p_i^+ - p_i^- \ (\forall i),$$

$$\sum_{j=1}^{n} (\lambda_j + \xi_j) y_{rj} + \Phi^{-1}(\alpha)\, \overline{\sigma}(q_r^+ + q_r^-) \ge y_{ro} \ (\forall r),$$

$$\sum_{j=1}^{n} (\lambda_j + \xi_j) b_{rj} - b_{ro} = q_r^+ - q_r^- \ (\forall r),$$

$$\sum_{j=1}^{n} (\xi_j - \gamma_j) z_{dj} - \Phi^{-1}(\alpha)\, \overline{\sigma}(\delta_d^+ + \delta_d^-) \le 0 \ (\forall d),$$

$$\sum_{j=1}^{n} (\xi_j - \gamma_j) c_{dj} = \delta_d^+ - \delta_d^- \ (\forall d),$$

$$p_i^+ p_i^- = 0 \ (\forall i),$$

$$q_r^+ q_r^- = 0 \ (\forall r),$$

$$\delta_d^+ \delta_d^- = 0 \ (\forall d),$$

$$\lambda_j, \gamma_j, \xi_j, \delta_d^+, \delta_d^-, q_r^+, q_r^-, p_i^+, p_i^- \ge 0 \ (\forall j, d, r, i).$$

Due to the existence of the constraints $p_i^+ p_i^- = 0, q_r^+ q_r^- = 0,$ and $\delta_d^+ \delta_d^- = 0$, model (5.46) is still non-linear. However, as we know, if a linear programming problem has an optimal solution, then it has an extreme optimal solution, implying that at least one of the variables $p_j^+ \ or \ p_j^-$ and $q_r^+ \ or \ q_r^-$ and $\delta_d^+ \ or \ \delta_d^-$ are equal to zero. As a result, if we use the simplex algorithm to solve this linear programming problem, we can find an extreme optimal solution and we can, hence, avoid these non-linear constraints, $p_i^+ p_i^- = 0, q_r^+ q_r^- = 0$, and $\delta_d^+ \delta_d^- = 0$ without any loss of information. Therefore, we need to solve, using the simplex algorithm, the following linear program:

$$e_{SCN}^{(\alpha)} = \text{Min}\theta \tag{5.47}$$

$$\text{s.t.} \quad \sum_{j=1}^{n} (\lambda_j + \gamma_j)x_{ij} - \Phi^{-1}(\alpha)\,\overline{\sigma}(p_i^+ + p_i^-) \le \theta x_{io} \ \ (\forall i),$$

$$\sum_{j=1}^{n} (\lambda_j + \gamma_j)a_{ij} - \theta a_{io} = p_i^+ - p_i^- \ \ (\forall i),$$

$$\sum_{j=1}^{n} (\lambda_j + \xi_j)y_{rj} + \Phi^{-1}(\alpha)\,\overline{\sigma}(q_r^+ + q_r^-) \ge y_{ro} \ \ (\forall r),$$

$$\sum_{j=1}^{n} (\lambda_j + \xi_j)b_{rj} - b_{ro} = q_r^+ - q_r^- \ \ (\forall r),$$

$$\sum_{j=1}^{n} (\xi_j - \gamma_j)z_{dj} - \Phi^{-1}(\alpha)\,\overline{\sigma}(\delta_d^+ + \delta_d^-) \le 0 \ \ (\forall d),$$

$$\sum_{j=1}^{n} (\xi_j - \gamma_j)c_{dj} = \delta_d^+ - \delta_d^- \ \ (\forall d),$$

$$\lambda_j, \gamma_j, \xi_j, \delta_d^+, \delta_d^-, q_r^+, q_r^-, p_i^+, p_i^- \ge 0.$$

Theorem 5.11 For each $\alpha' < \alpha < 0.5$, we have $e_{SCN}^{(\alpha')} \ge e_{SCN}^{(\alpha)}$.

Proof Suppose $(\theta^*, \lambda^*, \gamma^*, \xi^*, \delta^{+*}, \delta^{-*}, q^{+*}, q^{-*}, p^{+*}, p^{-*})$ is an optimal solution to model (5.47) when DMU_o is under evaluation at the confidence level of α. Since $\Phi^{-1}(\alpha)$ is a strictly increasing function, $\Phi^{-1}(\alpha') < \Phi^{-1}(\alpha)$ and we, therefore, have

$$\sum_{j=1}^{n} \left(\lambda_j^* + \gamma_j^*\right)x_{ij} - \Phi^{-1}(\alpha')\,\overline{\sigma}(p_i^{+*} + p_i^{-*})$$

$$\le \sum_{j=1}^{n} \left(\lambda_j^* + \gamma_j^*\right)x_{ij} - \Phi^{-1}(\alpha)\,\overline{\sigma}(p_i^{+*} + p_i^{-*}) \le \theta^* x_{io},$$

$$\sum_{j=1}^{n} \left(\lambda_j^* + \xi_j^*\right)y_{rj} + \Phi^{-1}(\alpha')\,\overline{\sigma}(q_r^{+*} + q_r^{-*})$$

$$\ge \sum_{j=1}^{n} \left(\lambda_j^* + \xi_j^*\right)y_{rj} + \Phi^{-1}(\alpha)\,\overline{\sigma}(q_r^{+*} + q_r^{-*}) \ge y_{ro},$$

$$\sum_{j=1}^{n}\left(\xi_j^* - \gamma_j^*\right) z_{dj} - \Phi^{-1}(\alpha')\,\overline{\sigma}\left(\delta_d^{+*} + \delta_d^{-*}\right)$$
$$\leq \sum_{j=1}^{n}\left(\xi_j^* - \gamma_j^*\right) z_{dj} - \Phi^{-1}(\alpha)\,\overline{\sigma}\left(\delta_d^{+*} + \delta_d^{-*}\right) \leq 0.$$

Hence, $(\theta^*, \lambda^*, \gamma^*, \xi^*, \delta^{+*}, \delta^{-*}, q^{+*}, q^{-*}, p^{+*}, p^{-*})$ is a feasible solution vector to model (5.47) when DMU_o under evaluation is at the confidence level of α'. The proof is hence complete due to that the model is a minimization problem. ∎

Corollary 5.3. If DMU_o is efficient under the confidence level of α, then it is also efficient under any confidence level α', which is less than α. Similarly, if DMU_o is inefficient at the confidence level α', then it is also inefficient at any confidence level α, which is greater than α'.

Theorem 5.12 Let $e_1^{(\alpha)}$ and $e_2^{(\alpha)}$ be the stochastic efficiency scores of the first and second stages in T_{stage1}^s and T_{stage2}^s, respectively. Then, $e_{SCN}^{(\alpha)} \leq e_1^{(\alpha)}$ and $e_{SCN}^{(\alpha)} \leq e_2^{(\alpha)}$.

Proof Suppose $\left(\lambda_1^{(1)*} + \gamma_1^{(1)*}, \lambda_2^{(1)*} + \gamma_2^{(1)*}, \ldots, \lambda_n^{(1)*} + \gamma_n^{(1)*}, e_1^{(\alpha)}\right)$ is an optimal solution to the stochastic input-oriented model of the first stage. Let $\lambda_j + \gamma_j = \lambda_j^{(1)*} + \gamma_j^{(1)*}$, $\lambda_j + \xi_j = \lambda_j^{(1)*} + \gamma_j^{(1)*}$, and $\gamma_j - \xi_j = 0$, for all $j = 1, \ldots, n$. As one can see, $\left(\lambda_1 + \gamma_1, \ldots, \lambda_n + \gamma_n, \lambda_1 + \xi_j, \ldots, \lambda_n + \xi_n, 0, \ldots, 0, e_1^{\alpha}\right)$ is also a feasible solution to model (5.47). Let the optimal value of model (5.47) be $e_{SCN}^{(\alpha)}$. Then, $e_{SCN}^{(\alpha)} \leq e_1^{(\alpha)}$. The proof of the second part is similar and is hence omitted. ∎

Corollary 5.4. If $e_1^{(\alpha)}$ and $e_2^{(\alpha)}$ are stochastic efficiency scores of the first and second stages in T_{stage1}^s and T_{stage2}^s, respectively, then, $e_{SCN}^{(\alpha)} \leq \frac{e_1^{(\alpha)} + e_2^{(\alpha)}}{2}$. Moreover, for any predetermined confidence level α, if the whole process is stochastically efficient in the stochastic centralized model, then each of its sub-processes is also stochastically efficient.

5.3.5 An Empirical Example

To illustrate the applicability of our proposed model, we now use the data set of 16 commercial banks used earlier by Zhou et al. (2017). We use model (5.47) to compute the efficiency scores of the overall processes at different confidence levels ($\alpha = 0.1, 0.2, \ldots, 0.5$), and present them in Table 5.7. As it can be seen, DMU 15 exhibits the best performance since it is stochastically efficient at all the confidence levels considered, and DMUs 11, 14, and 16 perform the worst at each of the confidence levels. The last column in Table 5.7 represents the stochastic efficiency scores at $\alpha = 0.5$, which are also the deterministic efficiency scores of the centralized model (5.39). On comparing the efficiency scores across different confidence levels, one can observe that the choice of a confidence level plays an important role in

Table 5.7 Efficiency scores from the stochastic centralized model ($\boldsymbol{e_{SCN}}$)

DMU	Confidence levels				
	$\alpha = 0.1$	$\alpha = 0.2$	$\alpha = 0.3$	$\alpha = 0.4$	$\alpha = 0.5$
1	0.444	0.399	0.366	0.334	0.304
2	0.329	0.312	0.297	0.280	0.262
3	0.254	0.220	0.194	0.169	0.145
4	0.392	0.353	0.327	0.303	0.279
5	0.370	0.326	0.290	0.256	0.222
6	0.349	0.331	0.313	0.294	0.268
7	0.312	0.297	0.284	0.268	0.251
8	0.441	0.416	0.392	0.364	0.330
9	0.433	0.394	0.366	0.340	0.310
10	0.513	0.463	0.422	0.383	0.342
11	0.202	0.166	0.152	0.139	0.126
12	0.413	0.384	0.359	0.334	0.307
13	0.421	0.395	0.372	0.348	0.323
14	0.172	0.158	0.149	0.140	0.130
15	1.000	1.000	1.000	1.000	1.000
16	0.225	0.197	0.174	0.152	0.127

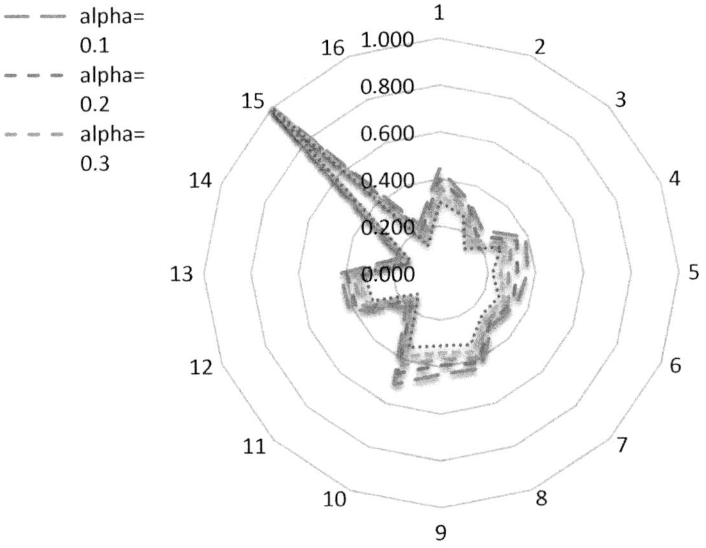

Fig. 5.5 The efficiency scores at different confidence levels

efficiency evaluation. To visualize this, Fig. 5.5 presents the efficiency scores at different confidence levels shown in Table 5.7. As expected, one can see a decrease in the efficiency scores of a DMU with an increase in α.

Table 5.8 presents the individual stochastic efficiencies of the two stages and the mean efficiency of the two stages across DMUs at the different confidence levels

Table 5.8 Efficiency scores and the mean efficiency of the two stages

	Confidence level														
	0.1			0.2			0.3			0.4			0.5		
DMU	e_1	e_2	e_{mean}	e_1	e_2	e_{mean}	e_1	e_2	e_{mean}	e_1	e_2	e_{mean}	e_1	e_2	e_{mean}
1	1.000	1.000	1.000	1.000	1.000	1.000	1.000	0.907	0.954	1.000	0.721	0.860	1.000	0.631	0.815
2	0.932	0.589	0.761	0.920	0.584	0.752	0.910	0.576	0.743	0.901	0.567	0.734	0.890	0.558	0.724
3	0.800	1.000	0.900	0.770	1.000	0.885	0.747	1.000	0.874	0.727	0.952	0.839	0.707	0.883	0.795
4	0.739	0.712	0.726	0.721	0.658	0.690	0.708	0.623	0.666	0.696	0.592	0.644	0.682	0.563	0.623
5	1.000	0.509	0.754	1.000	0.461	0.730	1.000	0.422	0.711	1.000	0.385	0.692	1.000	0.346	0.673
6	0.893	1.000	0.946	0.875	1.000	0.937	0.862	1.000	0.931	0.854	1.000	0.927	0.848	1.000	0.924
7	1.000	0.352	0.676	1.000	0.345	0.672	0.982	0.338	0.660	0.948	0.332	0.640	0.919	0.326	0.622
8	1.000	1.000	1.000	1.000	0.992	0.996	1.000	0.965	0.982	1.000	0.939	0.969	1.000	0.910	0.955
9	1.000	0.934	0.967	0.963	0.930	0.946	0.940	0.928	0.934	0.921	0.917	0.919	0.899	0.908	0.904
10	1.000	0.513	0.756	1.000	0.463	0.732	1.000	0.422	0.711	1.000	0.383	0.691	1.000	0.342	0.671
11	0.910	0.355	0.633	0.905	0.324	0.615	0.902	0.322	0.612	0.894	0.320	0.607	0.868	0.318	0.593
12	1.000	0.413	0.707	1.000	0.384	0.692	1.000	0.359	0.679	1.000	0.334	0.667	1.000	0.307	0.654
13	1.000	0.421	0.710	1.000	0.395	0.698	1.000	0.372	0.686	1.000	0.348	0.674	1.000	0.323	0.661
14	1.000	0.262	0.631	1.000	0.242	0.621	0.930	0.225	0.577	0.881	0.208	0.545	0.843	0.190	0.517
15	1.000	1.000	1.000	1.000	1.000	1.000	1.000	1.000	1.000	1.000	1.000	1.000	1.000	1.000	1.000
16	0.889	0.540	0.714	0.872	0.466	0.669	0.836	0.404	0.620	0.805	0.348	0.577	0.778	0.291	0.534

considered. Tables 5.7 and 5.8 reveal that, as expected, at each confidence level, the stochastic centralized efficiency of a DMU (e_{SCN}) is no more than the mean efficiency score (e_{mean}). For example, since DMU 15 is stochastically efficient in the centralized model, it is also efficient in both the stages. However, the converse is not necessarily true. For example, DMUs 1 and 8 are stochastically efficient in both stages at confidence level $\alpha = 0.1$, However, their overall efficiencies are not stochastically efficient. It is to be noted that in a deterministic centralized model, based on the definitions of the efficiency of the overall system and its stages, one can expect that the stage-specific efficiencies guarantee the efficiency of the overall system. In the stochastic case, however, this is not necessarily true.

References

Azadi, M., Jafarian, M., Saen, R. F., & Mirhedayatian, S. M. (2015). A new fuzzy DEA model for evaluation of efficiency and effectiveness of suppliers in sustainable supply chain management context. *Computers & Operations Research, 54*, 274–285.

Banker, R. D., & Natarajan, R. (2008). Evaluating contextual variables affecting productivity using data envelopment analysis. *Operations Research, 56*(1), 48–58.

Banker, R., Natarajan, R., & Zhang, D. (2019). Two-stage estimation of the impact of contextual variables in stochastic frontier production function models using data envelopment analysis: Second stage OLS versus bootstrap approaches. *European Journal of Operational Research, 278*(2), 368–384.

Charnes, A., & Cooper, W. W. (1962). Programming with linear fractional functionals. *Naval Research Logistics (NRL), 9*(3–4), 181–186.

Charnes, A., & Cooper, W. W. (1963). Deterministic equivalents for optimizing and satisficing under chance constraints. *Operations Research, 11*(1), 18–39.

Charnes, A., & Cooper, W. W. (1961). *Management models and industrial applications of linear programming* (No. 339.23 C4).

Charnes, A., & Cooper, W. W. (1977). Goal programming and multiple objective optimizations: Part 1. *European Journal of Operational Research, 1*(1), 39–54.

Cooper, W. W., Huang, Z., Lelas, V., Li, S. X., & Olesen, O. B. (1998). Chance constrained programming formulations for stochastic characterizations of efficiency and dominance in DEA. *Journal of Productivity Analysis, 9*(1), 53–79.

Cooper, W., Huang, Z., & Li, S. X. (1996). Chapter 13. Satisficing DEA models under chance constraints. *Annals of Operations Research, 66*(4), 279–295.

Dotoli, M., Epicoco, N., Falagario, M., & Sciancalepore, F. (2015). A cross-efficiency fuzzy data envelopment analysis technique for performance evaluation of decision-making units under uncertainty. *Computers & Industrial Engineering, 79*, 103–114.

Farrell, M. J. (1957). The measurement of productive efficiency. *Journal of the Royal Statistical Society: Series A (General), 120*(3), 253–281.

Hahn, G. J., Sens, T., Decouttere, C., & Vandaele, N. J. (2016). A multi-criteria approach to robust outsourcing decision-making in stochastic manufacturing systems. *Computers & Industrial Engineering, 98*, 275–288.

Huang, Z., & Li, S. X. (1996). Dominance stochastic models in data envelopment analysis. *European Journal of Operational Research, 95*(2), 390–403.

Izadikhah, M., & Saen, R. F. (2018). Assessing sustainability of supply chains by chance-constrained two-stage DEA model in the presence of undesirable factors. *Computers & Operations Research, 100*, 343–367.

Kahane, Y. (1977). Determination of the product mix and the business policy of an insurance company—a portfolio approach. *Management Science, 23*(10), 1060–1069.
Kao, C., & Hwang, S. N. (2011). Decomposition of technical and scale efficiencies in two-stage production systems. *European Journal of Operational Research, 211*(3), 515–519.
Kao, C., & Hwang, S.-N. (2008). Efficiency decomposition in two-stage data envelopment analysis: An application to non-life insurance companies in Taiwan. *European Journal of Operational Research, 185*(1), 418–429.
Kao, C., & Liu, S.-T. (2011). Efficiencies of two-stage systems with fuzzy data. *Fuzzy Sets and Systems, 176*(1), 20–35.
Khalili-Damghani, K., & Shahmir, Z. (2015). Uncertain network data envelopment analysis with undesirable outputs to evaluate the efficiency of electricity power production and distribution processes. *Computers & Industrial Engineering, 88*, 131–150.
Liang, L., Cook, W. D., & Zhu, J. (2008). DEA models for two-stage processes: Game approach and efficiency decomposition. *Naval Research Logistics (NRL), 55*(7), 643–653.
Liang, L., Yang, F., Cook, W. D., & Zhu, J. (2006). DEA models for supply chain efficiency evaluation. *Annals of Operations Research, 145*(1), 35–49.
Liu, S.-T. (2014). Restricting weight flexibility in fuzzy two-stage DEA. *Computers & Industrial Engineering, 74*, 149–160.
Lozano, S. (2014). Process efficiency of two-stage systems with fuzzy data. *Fuzzy Sets and Systems, 243*, 36–49.
Lu, C.-C. (2015). Robust data envelopment analysis approaches for evaluating algorithmic performance. *Computers & Industrial Engineering, 81*, 78–89.
Maghbouli, M., Amirteimoori, A., & Kordrostami, S. (2014). Two-stage network structures with undesirable outputs: A DEA based approach. *Measurement, 48*, 109–118.
Olesen, O. B., & Petersen, N. C. (2000). Foundation of chance constrained data envelopment analysis for pareto-Koopmann efficient production possibility sets. In *International DEA Symposium 2000, Measurement and Improvement in the 21st Century* (pp. 313–349). The University of Queensland.
Paradi, J. C., & Zhu, H. (2013). A survey on bank branch efficiency and performance research with data envelopment analysis. *Omega, 41*(1), 61–79.
Sahoo, B. K., Zhu, J., & Tone, K. (2014a). Decomposing efficiency and returns to scale in two-stage network systems. In *Data envelopment analysis* (pp. 137–164). Springer.
Sahoo, B. K., Zhu, J., Tone, K., & Klemen, B. M. (2014b). Decomposing technical efficiency and scale elasticity in two-stage network DEA. *European Journal of Operational Research, 233*(3), 584–594.
Seiford, L. M., & Zhu, J. (1999). Profitability and marketability of the top 55 US commercial banks. *Management Science, 45*(9), 1270–1288.
Sharpe, W. F. (1963). A simplified model for portfolio analysis. *Management Science, 9*(2), 277–293.
Simar, L., & Wilson, P. W. (2007). Estimation and inference in two-stage, semi-parametric models of production processes. *Journal of Econometrics, 136*(1), 31–64.
Wei, G., Chen, J., & Wang, J. (2014). Stochastic efficiency analysis with a reliability consideration. *Omega, 48*, 1–9.
Wei, Q., & Chang, T.-S. (2011). Optimal system design series-network DEA models. *Journal of the Operational Research Society, 62*(6), 1109–1119.
Yang, F., Wu, D., Liang, L., Bi, G., & Wu, D. D. (2011). Supply chain DEA: Production possibility set and performance evaluation model. *Annals of Operations Research, 185*(1), 195–211.
Yu, Y., & Shi, Q. (2014). Two-stage DEA model with additional input in the second stage and part of intermediate products as final output. *Expert Systems with Applications, 41*(15), 6570–6574.
Zhou, Z., Lin, L., Xiao, H., Ma, C., & Wu, S. (2017). Stochastic network DEA models for two-stage systems under the centralized control organization mechanism. *Computers & Industrial Engineering, 110*, 404–412.

Chapter 6
Stochastic Scale Elasticity

To analyze the performance of firms (decision-making units, DMUs) in the literature, several economic concepts such as economies of scale (returns to scale, RTS), economies of scope, marginal rates of technical substitutions, etc., have been used. Banker et al. (2004) studied RTS in different DEA models. In this chapter, however, we concentrate on determining and measuring RTS. Scale elasticity (SE) is a quantitative measure of the RTS characterization of the firms operating on the production frontier, which is used to determine improvement or deterioration in their productivities by resizing their scales of operation.

Both qualitative and quantitative approaches are used to evaluate the RTS characterizations of firms in the DEA literature. While the former deals with the identification of the types of RTS (*i.e.*, increasing, decreasing, or constant), the latter deals with the computation of SE. Most of the studies that apply bespoke computational methods to evaluate scale economies are confined to specific DEA technologies under the CRS specification of Charnes et al. (1978) and the VRS specification of Banker et al. (1984). However, there is a large class of polyhedral technologies that include the CRS and VRS technologies with production trade-offs and weight restrictions (Atici & Podinovski, 2015; Podinovski, 2004a, 2007, 2015, 2016; Tone, 2001), the weakly and managerially disposable technologies (Kuosmanen, 2005; Kuosmanen & Podinovski, 2009; Sueyoshi & Goto, 2012), the hybrid RTS technology (Podinovski, 2004b), the technology with multiple component processes (Cherchye et al., 2013, 2015, 2016; Cook & Zhu, 2011), the network technologies (Sahoo et al., 2014a, b), etc.

The SE evaluation has also been made using the cost-efficiency (CE) program of Färe et al. (1985) (Sueyoshi, 1997; Tone & Sahoo, 2005, 2006). The use of this program requires that input prices are exogenously given and are measured with full certainty. In real-life situations, however, input prices are not exogenous, but they vary by the actions of firms (Sahoo & Tone, 2013). Also, firms often face ex-ante

A. Amirteimoori et al., *Stochastic Benchmarking*, International Series in Operations Research & Management Science 317,
https://doi.org/10.1007/978-3-030-89869-4_6

price uncertainty[1] in making their production decisions (Camanho & Dyson, 2005). Furthermore, when input prices are synthetically constructed, they represent average prices. Since decisions are taken at the margin, the measure of allocative efficiency based on average prices can be distorted (Fukuyama & Weber, 2008).

To set up value-based measures of efficiency and SE, we first formulate, using the chance-constrained (CC) programming method, the stochastic version of the BCC program, which is essentially a quadratic program. Second, we set up the deterministic equivalent of this stochastic quadratic program. Third, with the help of the single-factor assumption, we convert this deterministic quadratic program into a linear program by using the goal programming theory of Charnes and Cooper (1977). Fourth, we solve this linear program to compute the value-based efficiency scores of firms at some predetermined tolerance levels of chance constraints. Finally, we set up the dual of this linear program to compute the value-based SE scores of firms at those tolerance levels.

6.1 A Value-Based Measure of Technical Efficiency and SE in the Deterministic Case

Consider a technology comprising of n firms with each one using m inputs to produce s outputs. Let $x_j = \left(x_{1j}, \ldots, x_{mj}\right)^T \in \mathbb{R}^m_{\geq 0}$ and $y_j = \left(y_{1j}, \ldots, y_{sj}\right)^T \in \mathbb{R}^s_{\geq 0}$ be, respectively, the input and output vectors of firm j, and let j be the index set of all firms, *i.e.* $j = 1, \ldots, n$. Following Banker et al. (1984), we set up the value-based technology set (T^V) as

$$T^V = \left\{ (x, y)^T \mid \sum_{j=1}^{n} \lambda_j x_{ij} \leq x_i \ (\forall i),\ \sum_{j=1}^{n} \lambda_j y_{rj} \geq y_r \ (\forall r),\ \sum_{j=1}^{n} \lambda_j = 1,\ \lambda \geq 0 \right\}. \tag{6.1}$$

To produce the output vector βy_o by firm o, its value-based measure of technical efficiency (TE) $\alpha(\beta)$ is defined as

$$\overline{\alpha}(\beta) = \text{Min}\{\alpha | (\alpha x_o, \beta y_o) \in T^V\}. \tag{6.2}$$

Note that β is a user-defined value to reflect the proportional output change, and the value-based technical efficiency function $\overline{\alpha}(\beta)$ is defined for all $\beta \in \left[0, \widehat{\beta}\right]$, where $\widehat{\beta}$ is the largest proportion of output vector y_o found in units in technology T^V.

[1]Even Farrell (1957) cautioned against the use of prices in determining the economic efficiency behavior of firms due to fluctuations in demand and prices (Sengupta, 2000).

Based on the definition in (6.2), $\overline{\alpha}(\beta)$ can be determined from the following linear program:

$$\overline{\alpha}(\beta) = \text{Min}\alpha$$
$$\text{s.t.} \quad \sum_{j=1}^{n} \lambda_j x_{ij} \leq \alpha x_{io}, \qquad i = 1, \ldots, m,$$
$$\sum_{j=1}^{n} \lambda_j y_{rj} \geq \beta y_{ro}, \qquad r = 1, \ldots, s \tag{6.3}$$
$$\sum_{j=1}^{n} \lambda_j = 1,$$
$$\lambda_j \geq 0 \ (\forall j), \quad \text{and } \alpha : \text{free}.$$

The dual formulation of program (6.3) can be set up as

$$\overline{\alpha}(\beta) = \text{Max } \beta \sum_{r=1}^{s} u_r y_{ro} + \omega_0$$
$$\text{s.t.} \quad \sum_{r=1}^{s} u_r \, y_{rj} - \sum_{i=1}^{m} v_i x_{ij} + \omega_0 \leq 0, \tag{6.4}$$
$$\sum_{i=1}^{m} v_i x_{io} = 1,$$
$$v, u \geq 0.$$

Based on program (6.4) for firm o, we derive its transformation function as

$$\psi(\overline{\alpha}(\beta) x_o, \beta y_o) = \beta \sum_{r=1}^{s} u_r \, y_{ro} - \overline{\alpha}(\beta) \sum_{i=1}^{m} v_i x_{io} + \omega_0$$
$$= 0. \tag{6.5}$$

Assume that the transformation function (6.5) is differentiable. Differentiating (6.5) with respect to β yields

$$\frac{\partial \psi(\overline{\alpha}(\beta) x_o, \beta y_o)}{\partial \beta} = \sum_{r=1}^{s} \frac{\partial \psi(\overline{\alpha}(\beta) x_o, \beta y_o)}{\partial (\beta y_{ro})} y_{ro} + \sum_{i=1}^{m} \frac{\partial \psi(\overline{\alpha}(\beta) x_o, \beta y_o)}{\partial (\overline{\alpha}(\beta) x_{io})} x_{io} \frac{\partial \alpha}{\partial \beta} = 0,$$

$$\Rightarrow \frac{\partial \overline{\alpha}(\beta)}{\partial \beta} = -\frac{\sum_{r=1}^{s} \frac{\partial \psi(\overline{\alpha}(\beta)x_o, \beta y_o)}{\partial(\beta y_{ro})} y_{ro}}{\sum_{i=1}^{m} \frac{\partial \psi(\overline{\alpha}(\beta)x_o, \beta y_o)}{\partial(\overline{\alpha}(\beta)x_{io})} x_{io}} = -\frac{\sum_{r=1}^{s} u_r y_{ro}}{-\sum_{i=1}^{m} v_i x_{io}} = \frac{\frac{\overline{\alpha}(\beta)-\omega_o}{\beta}}{\sum_{i=1}^{m} v_i x_{io}} = \frac{\frac{\overline{\alpha}(\beta)-\omega_o}{\beta}}{1}$$

$$= \frac{\overline{\alpha}(\beta) - \omega_o}{\beta}.$$

Following Li (1995a) and Banker et al. (2004), we define the value-based measure of SE of firm o, $\varepsilon(\overline{\alpha}(\beta)x_o, \beta y_o)$, as the ratio of its marginal utilization of input $\frac{\partial \overline{\alpha}(\beta)}{\partial \beta}$ to its average utilization of input $\frac{\overline{\alpha}(\beta)}{\beta}$, that is

$$\varepsilon(\overline{\alpha}(\beta)x_o, \beta y_o) = \frac{\partial \overline{\alpha}(\beta)}{\partial \beta} \cdot \frac{\beta}{\overline{\alpha}(\beta)} = \frac{\overline{\alpha}(\beta) - \omega_o}{\beta} \cdot \frac{\beta}{\overline{\alpha}(\beta)} = 1 - \frac{\omega_o}{\overline{\alpha}(\beta)}. \tag{6.6}$$

If the firm o is technically efficient, we have $\beta = \overline{\alpha}(\beta) = 1$ and $\varepsilon(x_o, y_o) = 1 - \omega_o$. In the immediately following section, we extend these results to the stochastic case.

6.2 A Value-Based Measure of TE and SE in the Stochastic Case

6.2.1 Stochastic Value-Based TE Measure

Let $\widetilde{x}_j = (\widetilde{x}_{1j}, \widetilde{x}_{2j}, \ldots, \widetilde{x}_{mj})^T \in \mathbb{R}^m_{\geq 0}$ and $\widetilde{y}_j = (\widetilde{y}_{1j}, \widetilde{y}_{2j}, \ldots, \widetilde{y}_{sj})^T \in \mathbb{R}^s_{\geq 0}$ be, respectively, the random input and output vectors of firm j. Further, let $E(\widetilde{x}_j) = x_j$ and $E(\widetilde{y}_j) = y_j$. Further, let all random inputs and outputs be jointly and normally distributed. The structure of our stochastic value-based technology set ($T^{V(S)}$) in the CC-programming setup is like the technology set in the BCC program. $T^{V(S)}$ is the union of confidence regions $D_j(1 - \gamma_j)$ with probability level $(1 - \gamma_j)$ $(j = 1, \ldots, n)$, where $D_j(1 - \gamma_j)$ is defined as

$$D_j(1 - \gamma_j) = \left\{ (\widehat{x}^T, \widehat{y}^T)^T \in \mathbb{R}^{m+s}_{+} \,\middle|\, \left[(\widehat{x} - x_j)^T, (\widehat{y} - y_j)^T \right] \Lambda_j^{-1} \left[(\widehat{x} - x_j)^T, (\widehat{y} - y_j)^T \right]^T \leq c_j^2 \right\}, \tag{6.7}$$

where Λ_j^{-1} represents the inverse of the variance–covariance matrix of $(\widetilde{x}_j, \widetilde{y}_j)$. Using the axioms of free disposability, ray unboundedness, convexity, minimal extrapolation, and inclusion of observations within confidence regions, $T^{V(S)}$ can be set up as

$$T^{V(S)} = \Big\{ \left(x^T, y^T\right)^T \in \mathbb{R}_+^{m+s} \,|\, \exists \left(\widehat{x}_j^T, \widehat{y}_j^T\right)^T \in D_j(1-\gamma_j) \text{ and } \lambda_j \geq 0,\ j \in \{1, \ldots, n\}$$
$$\text{such that } \sum_{j=1}^{n} \lambda_j \widehat{x}_{ij} \leq x_i\ (\forall i), \text{and } \sum_{j=1}^{n} \lambda_j \widehat{y}_{rj} \geq y_r\ (\forall r), \sum_{j=1}^{n} \lambda_j = 1 \Big\}, \tag{6.8}$$

in which c_j is determined by $\mathbb{P}\left(\chi^2_{M+S} \leq c_j^2\right) = 1 - \gamma_j$, and χ^2_{M+S} is the Chi-square random variable with $m + s$ degrees of freedom. See Olesen and Petersen (2000) for details.

Inspired by the deterministic value-based TE measure, $\overline{\alpha}(\beta)$ in program (6.3), we define its stochastic counterpart as the greatest possible radial contraction of $\widetilde{c}_o$ possible in $T^{V(S)}$ to produce $\beta\widetilde{y}_o$. Mathematically, the stochastic value based TE measure of firm o can be computed by solving program (6.9), where β is a predefined parameter.

$$\overline{\alpha}(\beta) = \text{Min}\alpha \tag{6.9}$$

$$\text{s.t.} \quad \mathbb{P}\left(\sum_{j=1}^{n} \lambda_j \widetilde{x}_{ij} \leq \alpha \widetilde{x}_{io}\right) \geq 1 - \gamma\ (\forall i), \tag{6.9.1}$$

$$\mathbb{P}\left(\sum_{j=1}^{n} \lambda_j \widetilde{y}_{rj} \geq \beta \widetilde{y}_{ro}\right) \geq 1 - \gamma\ (\forall r), \tag{6.9.2}$$

$$\sum_{j=1}^{n} \lambda_j = 1,$$

$$\lambda_j \geq 0\ (\forall j), \quad \text{and} \quad \alpha : \text{free}.$$

In program (6.9), we have used the "E-model" form of the marginal chance-constrained DEA program, which was introduced by Cooper et al. (2002) with the help of the BCC program by Banker et al. (1984). Therefore, we have used two separate probability constraints: one for the input cost and the other for the set of outputs.

The objective of program (6.9) is to measure the stochastic value-based TE of any firm o ($o = 1, 2, \ldots, n$) since its inputs and outputs are all assumed to be stochastic. This is done by minimizing the contraction factor α under some constraints concerning inputs and outputs. On the input side, there is a chance constraint, *i.e.* the best-practice (minimum) input must not exceed the observed input more than $\gamma\%$ of the time. As regards the output side, there are chance constraints according to which each observed output must not exceed its maximum more than $\gamma\%$ of the time. Finally, there is a convexity constraint, *i.e.* the sum of intensity coefficients is equal to 1. Note that γ is interpreted as the tolerance level of the chance constraint.

In the stochastic efficiency program (6.9), the random inputs and outputs of firm j are expressed in the following forms:

$$\widetilde{x}_{ij} = x_{ij} + a_{ij}\varepsilon_{ij}\ (\forall i) \quad \text{and} \quad \widetilde{y}_{rj} = y_{rj} + b_{rj}\eta_{rj}\ (\forall r), \tag{6.10}$$

where the error terms, $a_{ij}\varepsilon_{ij}$ and $b_{rj}\eta_{rj}$, represent the symmetric disturbances of inputs and outputs, respectively, which arise due to factors that are entirely outside of the firm's control. Further, ε_{ij} $(\forall i)$ and η_{rj} $(\forall r)$ are assumed to follow a multivariate normal distribution, with zero means $(E(\varepsilon_{ij}) = E(\eta_{rj}) = 0\ (\forall i, r))$ and unity variances $(Var(\varepsilon_{ij}) = Var(\eta_{rj}) = 1\ (\forall i, r))$. Then, for each firm j $(j = 1, \ldots, n)$, $\widetilde{x}_{ij}\tilde{N}\left(x_{ij}, a_{ij}^2\right)$ and $\widetilde{y}_{rj}\tilde{N}\left(y_{rj}, b_{rj}^2\right)$, where $E\left(\widetilde{x}_{ij}\right) = x_{ij}$ $(\forall i)$, $E\left(\widetilde{y}_{rj}\right) = y_{ij}$ $(\forall r)$, $Var\left(\widetilde{x}_{ij}\right) = a_{ij}^2$ $(\forall i)$, and $Var\left(\widetilde{y}_{rj}\right) = b_{rj}^2$ $(\forall r)$.

Note that the firm-specific symmetric disturbance terms, ε_{ij} $(\forall i)$ and η_{rj} $(\forall r)$, allow the technology set and its resulting efficiency frontier to vary arbitrarily across firms and capture the presence of measurement and specification errors, if any, in the data. It is also important to note that these errors vary with the level of confidence specified by the user.

Using the technique CC-programming theory, Land et al. (1993) proposed a method to transform the stochastic efficiency program (6.9) into a non-linear programming (NLP) problem. However, solving this NLP problem requires information on a substantial number of parameters in a covariance matrix between the components of the inputs and outputs. Therefore, to reduce the computational time involved in the estimation of such parameters, Cooper et al. (1996) suggested a linearization method using one simplifying assumption that the components of the inputs and outputs are all related only through a common relationship with a single factor, which has long been used in economics (Huang & Li, 1996, 2001; Kahane, 1977; Li, 1995a, b, 1998; Li & Huang, 1996; Sharpe, 1963). Following Cooper et al. (1996), we have also used this single-factor assumption in our proposed stochastic approach. That is, $\widetilde{x}_{ij} = x_{ij} + a_{ij}\varepsilon$ $(\forall i)$ and $\widetilde{y}_{rj} = y_{rj} + b_{rj}\varepsilon$ $(\forall r)$, where ε is a standard normal random variable with mean $E(\varepsilon) = 0$ and a constant (finite) standard deviation $\sigma(\varepsilon)$.

Note that the use of the single-factor assumption for linearization does not come free. This simplifying assumption, as Kao and Liu (2019) rightly pointed out, requires rigid assumptions in the resulting linear program that the correlation coefficient between any two stochastic inputs/outputs and between input and output, for any firm j, is always 1, *i.e.*, $\rho\left(\widetilde{x}_{pj}, \widetilde{x}_{qj}\right) = 1$ $(p \neq q)$, $\rho\left(\widetilde{y}_{rj}, \widetilde{y}_{sj}\right) = 1$ $(r \neq s)$, and $\rho\left(\widetilde{x}_{pj}, \widetilde{y}_{sj}\right) = 1$, which may not be easy to meet in practical applications. However, we believe that this convenient unit correlation coefficient assumption is important as it helps to convert an NLP problem into an LP problem, which not only removes the inconveniences associated with solving an NLP problem but also helps to compute the SE behavior of firms using its dual, as we will show later.

Using the single-factor assumption, for each firm j $(j = 1, \ldots, n)$, we express constraint (6.9.1) as

$$\widetilde{h} = \sum_{j=1}^{n} \lambda_j \widetilde{x}_{ij} - \alpha \widetilde{x}_{io}$$

$$= \sum_{j=1}^{n} \lambda_j \left(x_{ij} + a_{ij}\,\widetilde{\varepsilon} \right) - \alpha \left(x_{io} + a_{io}\,\widetilde{\varepsilon} \right)$$

$$= \sum_{j=1}^{n} \lambda_j x_{ij} - \alpha x_{io} + \widetilde{\varepsilon} \left(\sum_{j=1}^{n} \lambda_j a_{ij} - \alpha a_{io} \right).$$

Hence, we can see that

$$\widetilde{h} \sim N\left(\sum_{j=1}^{n} \lambda_j x_{ij} - \alpha x_{io}, \sigma^2(\varepsilon) \left(\sum_{j=1}^{n} \lambda_j a_{ij} - \alpha a_{io} \right)^2 \right).$$

Consider now a new variable $\widetilde{z} = \frac{\widetilde{h} - E\left(\widetilde{h}\right)}{\sigma_{\widetilde{h}}}$ that follows a standard normal distribution. As per the Central Limit Theorem, we have

$$\mathbb{P}\left(\widetilde{z} \leq \frac{-\left(\sum_{j=1}^{n} \lambda_j x_{ij} - \alpha x_{io} \right)}{\sigma(\varepsilon) \left| \sum_{j=1}^{n} \lambda_j a_{ij} - \alpha a_{io} \right|} \right) \geq 1 - \gamma.$$

Then,

$$\Phi\left(\frac{-\left(\sum_{j=1}^{n} \lambda_j x_{ij} - \alpha x_{io} \right)}{\sigma(\varepsilon) \left| \sum_{j=1}^{n} \lambda_j a_{ij} - \alpha a_{io} \right|} \right) \geq 1 - \gamma.$$

Therefore, using the property of invertibility of $\Phi(\cdot)$, we have

$$\frac{-\left(\sum_{j=1}^{n} \lambda_j x_{ij} - \alpha x_{io} \right)}{\sigma(\varepsilon) \left| \sum_{j=1}^{n} \lambda_j a_{ij} - \alpha a_{io} \right|} \geq \Phi^{-1}(1 - \gamma).$$

Consequently, the deterministic form of constraint (6.9.1) is expressed as

$$\sum_{j=1}^{n} \lambda_j x_{ij} - \Phi^{-1}(\gamma)\sigma(\varepsilon)\left|\sum_{j=1}^{n} \lambda_j a_{ij} - \alpha a_{io}\right| \leq \alpha x_{io} \ (\forall i).$$

Similarly, constraint (6.9.2) can be expressed as

$$\sum_{j=1}^{n} \lambda_j y_{rj} + \Phi^{-1}(\gamma)\sigma(\varepsilon)\left|\sum_{j=1}^{n} \lambda_j b_{rj} - \beta b_{ro}\right| \geq \beta y_{ro} \ (\forall r).$$

Thus, we set up the deterministic version of program (6.9) as

$$\begin{aligned} \overline{\alpha}(\beta) &= \text{Min}\alpha \\ \text{s.t.} \ &\sum_{j=1}^{n} \lambda_j x_{ij} - \Phi^{-1}(\gamma)\sigma(\varepsilon)\left|\sum_{j=1}^{n} \lambda_j a_{ij} - \alpha a_{io}\right| \leq \alpha x_{io}, \quad i = 1, \ldots, m, \\ &\sum_{j=1}^{n} \lambda_j y_{rj} + \Phi^{-1}(\gamma)\sigma(\varepsilon)\left|\sum_{j=1}^{n} \lambda_j b_{rj} - \beta b_{ro}\right| \geq \beta y_{ro}, \ r = 1, \ldots, s \\ &\sum_{j=1}^{n} \lambda_j = 1, \\ &\lambda_j \geq 0 \ (\forall j), \quad \text{and} \ \alpha : \text{free}. \end{aligned} \tag{6.11}$$

Program (6.11) is non-linear due to the existence of the absolute function. Using the goal programming theory by Charnes and Cooper (1977), we transform (6.11) into a quadratic program. To proceed, we use the following transformations:

$$\left|\sum_{j=1}^{n} \lambda_j a_{ij} - \alpha a_{io}\right| = p_i^+ + p_i^- \ (\forall i),$$

$$\sum_{j=1}^{n} \lambda_j a_{ij} - \alpha a_{io} = p_i^+ - p_i^- \ (\forall i),$$

$$p_i^+ p_i^- = 0 \ (\forall i),$$

$$\left|\sum_{j=1}^{n} \lambda_j b_{rj} - \beta b_{ro}\right| = q_r^+ + q_r^- \ (\forall r),$$

$$\sum_{j=1}^{n} \lambda_j b_{rj} - \beta b_{ro} = q_r^+ - q_r^- \ (\forall r),$$

$$q_r^+ q_r^- = 0 \ (\forall r).$$

Note that the existence of two constraints, $p_i^+ p_i^- = 0$ and $q_r^+ q_r^- = 0$, in our suggested transformations makes program (6.11) to be non-linear. However, as we know, if an LP problem exhibits an optimal solution vector, then it also has an extreme optimal solution vector in which at least one of the variables from (p_i^+, p_i^-) and (q_r^+, q_r^-) is each zero. Hence, the non-linear constraints $p_i^+ p_i^- = 0$ and $q_r^+ q_r^- = 0$ can be safely ignored. Consequently, one can solve, using the simplex algorithm, program (6.11) to find the extreme optimal solutions.

Using the above transformations and notations, program (6.11) is converted into the following program:

$$\begin{aligned}
&\overline{\alpha}(\beta) = \text{Min}\alpha \\
\text{s.t.} \quad &\sum_{j=1}^{n} \lambda_j x_{ij} - \Phi^{-1}(\gamma)\sigma(\varepsilon)\left(p_i^+ + p_i^-\right) \leq \alpha x_{io} \ (\forall i), \\
&\sum_{j=1}^{n} \lambda_j a_{ij} - \alpha a_{io} = p_i^+ - p_i^- \ (\forall i), \\
&\sum_{j=1}^{n} \lambda_j y_{rj} + \Phi^{-1}(\gamma)\sigma(\varepsilon)\left(q_r^+ + q_r^-\right) \geq \beta y_{ro} \ (\forall r), \\
&\sum_{j=1}^{n} \lambda_j b_{rj} - \beta b_{ro} = q_r^+ - q_r^- \ (\forall r), \\
&\sum_{j=1}^{n} \lambda_j = 1, \\
&\lambda, p^+, p^-, q^+, q^- \geq 0, \quad \text{and } \alpha : \text{free}.
\end{aligned} \tag{6.12}$$

6.2.2 *Stochastic Value-Based SE Measure*

To compute the value-based measure of SE of firm o, we set up the dual of program (6.12) as

$$
\begin{aligned}
\overline{\alpha}(\beta) = \text{Max } \beta & \left(\sum_{r=1}^{S} (u_r y_{ro} + \mu_r b_{ro}) \right) + \omega_0 \\
\text{s.t.} \quad & \sum_{r=1}^{S} \left(u_r y_{rj} + \mu_r b_{rj}\right) - \sum_{i=1}^{m} \left(v_i x_{ij} - \vartheta_i a_{ij}\right) + \omega_0 \leq 0 \quad (\forall j), \\
& \sum_{i=1}^{m} (v_i x_{io} - \vartheta_i a_{io}) = 1, \\
& \Phi^{-1}(\gamma)\sigma(\varepsilon)v_i - \vartheta_i \leq 0, \\
& \Phi^{-1}(\gamma)\sigma(\varepsilon)v_i + \vartheta_i \leq 0, \\
& \Phi^{-1}(\gamma)\sigma(\varepsilon)u_r - \mu_r \leq 0, \\
& \Phi^{-1}(\gamma)\sigma(\varepsilon)u_r + \mu_r \leq 0, \\
& v, u \geq 0, \qquad \vartheta, \mu, \omega_0 : \text{free}.
\end{aligned}
\tag{6.13}
$$

For each firm $o \in \{1, \ldots, n\}$, consider its transformation function:

$$
\begin{aligned}
\psi(\overline{\alpha}(\beta)x_o, \overline{\alpha}(\beta)a_o, \beta y_o, \beta b_o) \equiv \beta \sum_{r=1}^{S} (u_r y_{ro} + \mu_r b_{ro}) - \overline{\alpha}(\beta) \sum_{i=1}^{m} \\
\times (v_i x_{io} - \vartheta_i a_{io}) + \omega_0 \\
= 0.
\end{aligned}
\tag{6.14}
$$

By differentiating (6.14) with respect to β, we have

$$
\begin{aligned}
\frac{\partial \psi(\cdot)}{\partial \beta} = & \sum_{r=1}^{s} \frac{\partial \psi(.)}{\partial(\beta y_{ro})} y_{ro} + \sum_{r=1}^{s} \frac{\partial \psi(\cdot)}{\partial(\beta b_{ro})} b_{ro} + \sum_{i=1}^{m} \frac{\partial \psi(.)}{\partial(\alpha x_{io})} x_{io} \frac{\partial \alpha}{\partial \beta} + \sum_{i=1}^{m} \\
& \times \frac{\partial \psi(.)}{\partial(\alpha a_{io})} a_{io} \frac{\partial \alpha}{\partial \beta} \\
= & \, 0,
\end{aligned}
$$

$$
\Rightarrow \frac{\partial \alpha}{\partial \beta} = - \frac{\sum_{r=1}^{s} \frac{\partial \psi(\cdot)}{\partial(\beta y_{ro})} y_{ro} + \sum_{r=1}^{s} \frac{\partial \psi(\cdot)}{\partial(\beta b_{ro})} b_{ro}}{\sum_{i=1}^{m} \frac{\partial \psi(\cdot)}{\partial(\alpha x_{io})} x_{io} + \sum_{i=1}^{m} \frac{\partial \psi(\cdot)}{\partial(\alpha a_{io})} a_{io}}
$$

$$
= - \frac{\sum_{r=1}^{s} u_r y_{ro} + \sum_{r=1}^{s} \mu_r b_{ro}}{-\sum_{i=1}^{m} v_i x_{io} + \sum_{i=1}^{m} \vartheta_i a_{io}}
$$

$$= \frac{\overline{\alpha}(\beta)\sum_{i=1}^{m}(v_i x_{io} - \vartheta_i a_{io}) - \omega_0}{\beta} = \frac{\overline{\alpha}(\beta) - \omega_0}{\beta}. \tag{6.15}$$

The (input-oriented) value-based measure of SE of firm o can then be derived as

$$\begin{aligned} \varepsilon(\gamma) &= \frac{\partial \overline{\alpha}(\beta)}{\partial \beta} \frac{\beta}{\overline{\alpha}(\beta)} = \frac{\overline{\alpha}(\beta) - \omega_0}{\beta} \frac{\beta}{\overline{\alpha}(\beta)} \\ &= \frac{\overline{\alpha}(\beta) - \omega_0}{\overline{\alpha}(\beta)}. \end{aligned} \tag{6.16}$$

As the DEA technologies are not smooth at vertices, we face multiple optimal values for ω_0. Therefore, to compute the maximum (minimum) value of ω_0 for firm o, we set up the following program:

$$\begin{aligned} \omega_0^+ (\omega_0^-) &= \text{Max (Min) } \omega_0 \\ \text{s.t.} \quad & \beta \left(\sum_{r=1}^{s} (u_r y_{ro} + \mu_r b_{ro}) \right) + \omega_0 = \overline{\alpha}(\beta), \\ \sum_{r=1}^{s} (u_r y_{rj} + \mu_r b_{rj}) - \sum_{i=1}^{m} & (v_i x_{ij} - \vartheta_i a_{ij}) + \omega_0 \le 0 \quad (\forall j \ne o), \\ & \sum_{i=1}^{m} (v_i x_{io} - \vartheta_i a_{io}) = 1, \\ & \Phi^{-1}(\gamma)\sigma(\varepsilon) v_i - \vartheta_i \le 0, \\ & \Phi^{-1}(\gamma)\sigma(\varepsilon) v_i + \vartheta_i \le 0, \\ & \Phi^{-1}(\gamma)\sigma(\varepsilon) u_r - \mu_r \le 0, \\ & \Phi^{-1}(\gamma)\sigma(\varepsilon) u_r + \mu_r \le 0, \\ & v, u \ge 0, \quad \vartheta, \mu, \omega_0 : \text{free}. \end{aligned} \tag{6.17}$$

Using the optimal values of program (6.17), the (input-oriented) right- and left-hand value-based SEs of firm o can be calculated, respectively, as

$$\varepsilon^+(\gamma) = \frac{\overline{\alpha}(\beta) - \omega_0^-}{\overline{\alpha}(\beta)} \quad \text{and} \quad \varepsilon^-(\gamma) = \frac{\overline{\alpha}(\beta) - \omega_0^+}{\overline{\alpha}(\beta)}. \tag{6.18}$$

We now present Theorem 6.1 to determine, using the formulae in (6.18), the RTS types at each confidence level, given the values of ω_0^- and ω_0^+.

Theorem 6.1 For every confidence level γ, the (input-oriented) RTS of firm o is (a) decreasing (DRS) (*i.e.*, $\varepsilon_I^-(\gamma) > 1$) if $\omega_0^+ < 0$, (b) constant (CRS) (*i.e.*, $\varepsilon_I^-(\gamma) \leq 1 \leq \varepsilon_I^+(\gamma)$) if $\omega_0^- \leq 0 \leq \omega_0^+$, and (c) increasing (IRS) (*i.e.*, $\varepsilon_I^+(\gamma) < 1$) if $\omega_0^- > 0$.

Proof Let v^*, u^*, ϑ^*, μ^*, and ω_0^* be the optimal solutions of the LP program (6.13) for the radial efficient firm $(\widetilde{x}_o, \widetilde{y}_o)$, which is defined according to (6.10) as $\widetilde{x}_{io} = x_{io} + a_{io}\varepsilon_{io}$ $(\forall i)$ and $\widetilde{y}_{ro} = y_{ro} + b_{ro}\eta_{ro}$ $(\forall r)$. So, $\sum_{r=1}^{S} \left(u_r^* y_{ro} + \mu_r^* b_{ro}\right) + \omega_0^* = \sum_{i=1}^{m} \times (v^* x_{io} - \vartheta^* a_{io}) = 1$. Furthermore, $\sum_{r=1}^{S} \left(u_r^* y_{rj} + \mu_r^* b_{rj}\right) - \sum_{i=1}^{m} \left(v^* x_{ij} - \vartheta^* a_{ij}\right) + \omega_0^* = 0$ is a supporting hyperplane to $T^{V(S)}$ in (6.8) and the corresponding value at $(\overline{\alpha}(\beta), \beta)$ is presented in (6.14). By differentiating (6.14) with respect to β, we get the following measure of SE: $\varepsilon(\gamma) = \frac{\overline{\alpha}(\beta) - \omega_o^*}{\overline{\alpha}(\beta)}$. One can then easily compute the right- and left-hand SEs as $\varepsilon^+(\gamma) = \frac{\overline{\alpha}(\beta) - \omega_o^{*-}}{\overline{\alpha}(\beta)}$ and $\varepsilon^-(\gamma) = \frac{\overline{\alpha}(\beta) - \omega_o^{*+}}{\overline{\alpha}(\beta)}$, respectively. It then immediately follows that:

(a) $\varepsilon^-(\gamma) > 1$ if $\omega_o^{*+} < 0$, and since $\omega_o^{*-} < \omega_o^{*+}$, then $\varepsilon^+(\gamma) > 1$. So, RTS is decreasing (DRS);
(b) if $\omega_o^{*-} < 0$, then $\varepsilon^+(\gamma) > 1$, and if $\omega_o^{*+} > 0$, then $\varepsilon^-(\gamma) < 1$. So, $\varepsilon^-(\gamma) \leq 1 \leq \varepsilon^+(\gamma)$ and RTS is constant (CRS); and
(c) if $\omega_o^{*-} > 0$, then $\varepsilon^+(\gamma) < 1$, and since $\omega_o^{*+} > \omega_o^{*-}$, then $\varepsilon^-(\gamma) < 1$. So, RTS is increasing (IRS). ∎

Regarding the relationship on RTS types between the deterministic and stochastic cases, we present Theorem 6.2.

Theorem 6.2 For a predetermined significance level of $\gamma = 0.5$, the RTS types obtained from the deterministic and stochastic programs are the same.

Proof To prove this, it suffices to show that for $\gamma = 0.5$, program (6.12) and program (6.3) are equivalents. To this end, since program (6.12) and program (6.11) are equivalent and $\Phi^{-1}(0.5) = 0$, then replacing it in the constraints of program (6.11) yields exactly program (6.3). ∎

In what follows, to illustrate the foregoing theoretical framework, we consider a hypothetical data set consisting of six firms (DMUs A-F) with one input and one output (Table 6.1).

Table 6.1 Hypothetical data set

DMUs	Input (x)	Output (y)
A	(1.5, 0.5)	(1, 0.51)
B	(4, 0.4)	(6, 0.6)
C	(6, 0.6)	(7, 0.7)
D	(2, 0.2)	(2.5, 2.5)
E	(4, 0.4)	(3.5, 0.5)
F	(7, 0.7)	(6.5, 0.1)

Note: For (a, b), a is the mean and b is the standard deviation

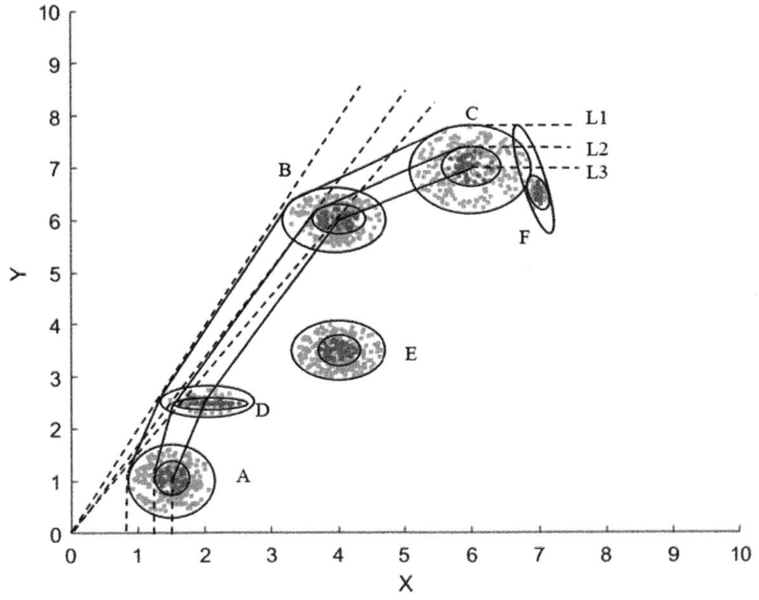

Fig. 6.1 DEA stochastic production frontiers under chance constraints at $\gamma = 0.5$, 0.3, and 0.1

Based on the data in Table 6.1, we construct the stochastic production frontiers (Fig. 6.1) by considering three different levels of tolerance ($\gamma = 0.1$, 0.3, and 0.5).[2] The piecewise linear line (L3) passing through the mean points of the centers of circles represents the production frontier at the tolerance level of 0.5 and, according to Theorem 6.2, this stochastic frontier is the same as the deterministic frontier. Likewise, the lines L2 and L1 represent the production frontiers at the tolerance levels of 0.3 and 0.1, respectively. The results of TE scores at these tolerance levels

[2]Cooper et al. (2000) suggested, in general, to consider the following tolerance level of the chance constraints: $0 < \gamma \leq 0.5$, for which the efficiency scores lie between 0 and 1. However, for data with large amounts of uncertainty, where $0.5 < \gamma < 1$, the efficiency frontier moves closer to the given observations, and many observations will automatically come with unity or close-to-unity efficiency scores. The results of the CC program may then be questioned. That is why most of the studies in the stochastic case have considered the values for γ between 0 and 0.5, *i.e.* $0 < \gamma \leq 0.5$.

are reported in Table 6.2. At each of these three tolerance levels, four firms A, B, C, and D are consistently found technically efficient. Out of the remaining two firms, F becomes efficient at the tolerance level of 0.1 but inefficient at the tolerance levels of 0.3 and 0.5. However, E is found inefficient at all three tolerance levels. This finding suggests that the efficiency status of a firm in the deterministic environment can be subject to change in the stochastic case.

Using the value-based technology (6.8) in the stochastic environment ($T^{V(S)}$), the input-oriented right- and left-hand SEs are computed. Table 6.2 presents the results.

At each of these three tolerance levels considered, out of four stochastically technically efficient firms (A, B, C, and D), B is CCR-efficient and, hence, exhibits CRS since its left-hand SE scores are all less than 1 and the right-hand SE scores are all greater than 1. D is CCR-efficient only at the tolerance levels of 0.1 and 0.3, and, hence, exhibits CRS at those tolerance levels, but IRS at the tolerance level of 0.5 (as its right-hand SE score is less than 1). Out of the remaining two stochastically technically efficient firms, while A exhibits IRS (as its right-hand SE score is less than 1), C exhibits DRS (as its left-hand SE score is greater than 1) at each of those three tolerance levels. Finally, as regards the RTS types of inefficient firms, since E's projection onto the frontier is located on the interior point of facet DB (see Fig. 6.1), its right- and left-hand SE values are all equal but all less than one, implying IRS. As it is evident from Table 6.2, the efficiency status of F is the mix, *i.e.* it is efficient at the tolerance level of 0.1, but inefficient at other tolerance levels. However, at all these levels of tolerance, it exhibits DRS since its left-hand SE values are all greater than one. Note that the finding that D exhibits IRS at the tolerance level of 0.5 but CRS at the other tolerance levels indicates that the RTS characterizations of firms in a deterministic technology can be subject to change in the stochastic case.

6.3 The Illustrative Empirical Application

We now turn to demonstrate an illustrative empirical application of our proposed stochastic efficiency program. To this aim, we employ the data set used by Ray et al. (2008) to analyze the production location decision of a manufacturing firm across 48 adjacent states within the United States.

6.3.1 The Data

The data set consists of five inputs, *i.e.*, production labor (x1), non-production labor (x2), energy (x3), materials (x4), and capital (x5), and one output, *i.e.*, the gross value of production (y). Production labor is measured as the number of hours worked. Non-production labor is measured by the number of non-production employees. Energy represents the deflated expenditure value of purchased fuels and electricity. Material input represents the total expenditure on materials, parts,

Table 6.2 Stochastic TE and SE scores of six firms

DMUs	$\gamma = 0.1$				$\gamma = 0.3$				$\gamma = 0.5$			
	TE_o	ε_1^-	ε_1^+	RTS	TE_o	ε_1^-	ε_1^+	RTS	TE_o	ε_1^-	ε_1^+	RTS
A	1	0	0.640	IRS	1	0	0.390	IRS	1	0	0.223	IRS
B	1	0.563	3.00	CRS	1	0.668	3.000	CRS	1	0.857	3.000	CRS
C	1	2.334	∞	DRS	1	2.334	∞	DRS	1	2.334	∞	DRS
D	1	0.072	5.201	CRS	1	0.244	1.523	CRS	1	0.417	0.714	IRS
E	0.745	0.414	0.414	IRS	0.713	0.534	0.534	IRS	0.642	0.778	0.778	IRS
F	1	3.385	∞	DRS	0.801	2.427	2.427	DRS	0.714	2.600	2.600	DRS

and containers. Capital is measured as the average of the beginning of the year and end of the year value of gross fixed assets. The inputs and output are all assumed to be random, with each one following a normal distribution with a known mean (presented in Table 6.3) and standard deviation (presented in Table 6.4).

6.3.2 *Stochastic Value-Based TE Scores*

Table 6.5 presents the results of the value-based TE scores in the stochastic environment for the three different confidence levels considered. As it can be seen, only 21 states are efficient at all three confidence levels. Also, it is interesting to observe that the inefficient states tend to experience a decline in their efficiency performance with an increase in γ.

6.3.3 *Stochastic Value-Based SE Scores*

The results of both the lower and upper bounds of the stochastic value-based SE scores for all the 48 states are presented in Tables 6.6 and 6.7 for $\beta = 0.99$ and $\beta = 1.01$, respectively.

As it is well known, scale elasticity is defined only for efficient units; for inefficient units, this computation can be made only at their input- or output-oriented projection points. In our case, however, we have considered their input-oriented projections. To interpret our results for $\beta = 0.99$ in Table 6.6, see states 2, 6, 7, 16, 19, 29, 30, 31, 37, 39, and 43 that are found efficient at all the three tolerance levels considered, their lower-bound SE score is less than 1, and their upper-bound value is more than 1, *i.e.*, $\varepsilon^- < 1 < \varepsilon^+$, implying that their RTS are constant (as per Theorem 6.1 (b)). For the efficient states 8, 24, and 26, the upper bound elasticity scores are all less than 1 (*i.e.*, $\varepsilon^+ < 1$) at all the three tolerance levels, implying that their RTS are increasing (as per Theorem 6.1 (c)); and for the efficient states such as 14, 25, and 38, since their lower-bound SE score is greater than 1 at all the three tolerance levels (*i.e.*, $\varepsilon^+ > 1$), they exhibit decreasing RTS (as per Theorem 6.1 (a)).

Notice that for some stochastically inefficient states such as 1, 3, 4, and 5, both lower- and upper-bound SEs are all equal at all the three tolerance levels, implying that their projections onto the boundary are all located on *interior* points of the facet. Out of 48 states, we find 18 states exhibiting DRS, 11 states CRS, and 10 states IRS, at all three tolerance levels. Two more interesting observations can be made. First, some states exhibit RTS of different types at different tolerance levels. For example, for $\beta = 0.99$ in Table 6.6, state 17 as an inefficient state at $\gamma = 0.1$ and 0.3 exhibits IRS. This means that its projections onto the efficient frontier are all located on points of the facet exhibiting IRS. However, the projection of this inefficient state at $\gamma = 0.5$ onto the efficient frontier exhibits DRS. Furthermore, state 28 at $\gamma = 0.1$ is both VRS- and CRS-efficient and, hence, exhibits CRS. However, this state is found

Table 6.3 Manufacturing random average inputs and output for individual states (per firm)

States	x_1	x_2	x_3	x_4	x_5	y	States	x_1	x_2	x_3	x_4	x_5	y
1	85.72	12.46	82.7	6.648	7.289	13.087	25	83.08	11.92	28.71	8.806	5.159	15.519
2	41.76	12.81	11.25	2.842	3.592	8.442	26	31.7	8	8.14	1.883	2.139	4.819
3	104.99	12.72	55.5	7.001	6.832	14.71	27	48.02	13.36	9.22	2.647	3.22	6.908
4	41.34	12.1	13.26	3.155	2.951	7.805	28	43.82	12.38	16.1	3.578	3.606	8.998
5	35.22	10.07	10.44	2.604	3.175	6.449	29	29.03	5.85	13.81	2.386	3.236	6.45
6	46.61	16.53	10.43	2.564	3.624	8.29	30	39.63	10.11	13.66	2.456	2.778	6.996
7	72.71	15.7	47.5	14.05	10.057	23.087	31	86.19	14.04	31.82	5.743	5.727	14.577
8	31.73	8.68	12.33	2.064	2.228	5.164	32	45.3	8.77	38.23	5.279	3.994	9.343
9	78.82	12.14	44.72	6.743	5.872	14.249	33	72.73	13.08	30.44	6.747	5.718	13.888
10	50.31	7.96	34.8	3.797	5.63	8.371	34	53.33	10.12	25.59	5.181	4.044	9.939
11	60.92	13.68	25.07	4.964	4.906	11.135	35	45.48	9.59	22.61	3.003	3.716	8.228
12	92.7	15.1	54.25	8.141	7.798	17.443	36	60.45	12.48	28.33	4.595	4.692	10.881
13	84.74	15.73	44.14	7.575	6.708	17.06	37	39.98	9.13	6.34	1.898	2.134	5.063
14	80.21	15.82	33.97	8.298	5.204	15.697	38	98.97	15.65	61.72	8.297	11.377	18.155
15	94.13	14.39	67.2	11.54	8.124	20.659	39	61.06	9.11	16.93	5.409	2.956	11.539
16	62.57	11.69	206.65	15.4	16.322	25.469	40	88.79	14.52	44.46	7.699	7.621	15.759
17	51.11	9.69	51.94	2.929	5.436	7.343	41	54.91	12.47	80.36	7.87	7.5	14.551
18	46.05	14.02	27.24	3.537	4.08	9.058	42	46.54	11.76	16.97	3.751	3.416	8.188
19	47.09	15.64	11.27	3.103	3.254	8.735	43	48.34	13.09	9.59	2.966	5.076	8.051
20	71.56	12.28	25.45	7.593	5.896	14.59	44	78.21	13.91	40.59	5.387	6.417	14.197
21	56.36	14.55	22.1	4.432	4.026	9.886	45	44.32	12.93	35.02	5.097	4.537	10.315
22	99.43	13.17	63.57	6.988	7.289	13.651	46	68.63	11.4	86.68	6.329	9.032	12.804
23	63.95	11.15	21.65	6.635	4.729	12.877	47	71.6	13.97	28.74	5.562	5.211	12.556
24	21.53	3.92	32.78	2.366	2.271	4.045	48	24.75	4.17	58.81	4.026	4.494	7.3

Table 6.4 The standard deviation of inputs and outputs for individual states in the United States

States	x_1	x_2	x_3	x_4	x_4	y	States	x_1	x_2	x_3	x_4	x_5	y
1	2.63	0.92	2.28	0.67	0.69	1.22	25	2.35	0.92	1.38	0.8	0.64	1.14
2	1.74	0.89	0.9	0.43	0.49	1.02	26	1.56	0.8	0.82	0.37	0.4	0.66
3	2.73	0.98	2.1	0.71	0.75	1.25	27	1.85	1	0.76	0.43	0.47	0.77
4	1.72	0.92	0.92	0.49	0.43	0.91	28	1.67	0.94	1.08	0.51	0.55	0.87
5	1.57	0.9	0.82	0.44	0.49	0.79	29	1.45	0.67	1.05	0.43	0.49	0.84
6	1.73	1.08	0.89	0.45	0.53	0.88	30	1.79	0.81	1.06	0.43	0.45	0.89
7	2.25	1.13	1.79	1.08	0.9	1.39	31	2.58	0.95	1.43	0.68	0.63	1.14
8	1.53	0.8	0.96	0.38	0.38	0.69	32	1.89	0.79	1.78	0.64	0.5	1.03
9	2.32	0.92	1.69	0.71	0.64	1.15	33	2.36	1	1.47	0.66	0.65	1.28
10	1.78	0.81	1.7	0.51	0.63	0.93	34	1.98	0.81	1.44	0.64	0.54	1.11
11	2.23	1	1.39	0.62	0.61	1.06	35	1.7	0.8	1.24	0.44	0.52	0.9
12	2.64	1.03	2	0.72	0.77	1.26	36	2.06	0.9	1.36	0.57	0.56	1.11
13	2.33	0.99	1.73	0.74	0.66	1.31	37	1.61	0.79	0.65	0.37	0.4	0.68
14	2.5	1.02	1.57	0.77	0.63	1.35	38	2.81	1.1	1.99	0.81	0.94	1.34
15	2.79	1.04	2.09	0.9	0.71	1.46	39	1.97	0.81	1.11	0.62	0.48	0.99
16	2.01	0.95	3.59	1.04	1.14	1.47	40	2.51	1.01	1.88	0.76	0.74	1.39
17	1.95	0.83	2.04	0.46	0.61	0.8	41	2.04	0.93	2.33	0.73	0.71	1.29
18	1.85	1.08	1.42	0.51	0.54	1.04	42	1.94	0.9	1.14	0.53	0.49	0.85
19	1.83	1.01	0.89	0.46	0.45	1.01	43	1.85	0.92	0.84	0.46	0.58	0.84
20	2.36	0.93	1.29	0.77	0.62	1.33	44	2.51	1.04	1.69	0.66	0.68	1.2
21	1.95	0.96	1.2	0.54	0.51	1.04	45	1.92	0.93	1.71	0.65	0.59	1.03
22	2.85	0.98	2.16	0.69	0.7	1.13	46	2.2	0.85	2.67	0.65	0.81	1.08
23	2.1	0.84	1.2	0.65	0.58	1.2	47	2.15	1.03	1.46	0.68	0.6	1.03
24	1.3	0.51	1.6	0.39	0.38	0.6	48	1.27	0.55	2.04	0.54	0.59	0.82

Table 6.5 Stochastic value-based TE scores of the states

States	$\gamma = 0.1$	$\gamma = 0.3$	$\gamma = 0.5$	States	$\gamma = 0.1$	$\gamma = 0.3$	$\gamma = 0.5$
1	0.886	0.876	0.868	25	1	1	1
2	1	1	1	26	1	1	1
3	0.958	0.952	0.948	27	0.957	0.948	0.942
4	1	0.999	0.995	28	1	1	0.991
5	0.998	0.994	0.990	29	1	1	1
6	1	1	1	30	1	1	1
7	1	1	1	31	1	1	1
8	1	1	1	32	0.947	0.935	0.926
9	0.965	0.962	0.960	33	0.961	0.949	0.941
10	0.905	0.902	0.899	34	0.900	0.889	0.881
11	0.912	0.908	0.906	35	1	0.999	0.993
12	1	0.996	0.992	36	0.941	0.936	0.931
13	1	1	1	37	1	1	1
14	1	1	1	38	1	1	1
15	1	1	1	39	1	1	1
16	1	1	1	40	0.948	0.938	0.930
17	0.916	0.903	0.892	41	0.969	0.961	0.956
18	0.943	0.940	0.939	42	0.905	0.898	0.893
19	1	1	1	43	1	1	1
20	1	1	0.998	44	1	1	1
21	0.894	0.890	0.886	45	0.981	0.973	0.967
22	0.871	0.866	0.862	46	0.914	0.908	0.904
23	0.979	0.969	0.958	47	0.937	0.929	0.923
24	1	1	1	48	1	1	1

inefficient at $\gamma = 0.5$. Hence, its projection onto the efficient frontier exhibits IRS. The SE results in the case of $\beta = 1.01$ in Table 6.7 can be interpreted analogously.

It can be observed from our results that different values of β and γ lead to varying resultson the RTS possibilities of a state. In such cases, the relevant question arises as to how an appropriate judgment for the RTS of a state can be made. As regards the possible values of β, as we have noted earlier, it is a user-defined value to reflect the proportional output change, and the value-based TE function $\overline{\alpha}(\beta)$ is defined for all $\beta \in \left[0, \widehat{\beta}\right]$, where $\widehat{\beta}$ is the largest proportion of the output vector y_o found in states in technology T^V. However, the scale elasticity of any radial technically efficient state (x_o, y_o) is well-defined at $\beta = \alpha(\beta) = 1$ as we determine its RTS status locally at the neighborhood of point (x_o, y_o), where a derivative is to be evaluated. Since scale elasticity is a one-sided concept, one can, therefore, determine a state's RTS type at its right or left by measuring the response to outputs by considering only 1% increase or decrease in inputs, *i.e.* $\beta = 1.01$ (right) and $\beta = 0.99$ (left) at a specific confidence level.

And, as regards the values of γ, it is also a predetermined parameter ($0 < \gamma \leq 0.5$), which makes the confidence level of each firm large or small, *i.e.* the confidence

Table 6.6 The stochastic value-based scale elasticity scores of 48 states ($\beta = 0.99$)

	$\gamma = 0.1$			$\gamma = 0.3$			$\gamma = 0.5$		
States	ε^-	ε^+	RTS	ε^-	ε^+	RTS	ε^-	ε^+	RTS
1	1.0406	1.0406	D	1.0406	1.0406	D	1.0410	1.0410	D
2	0.5432	2.1056	C	0.5677	1.8209	C	0.5848	1.6184	C
3	1.0636	1.0636	D	1.0480	1.0480	D	1.0459	1.0459	D
4	0.6892	0.8130	I	0.7440	0.7440	I	0.7432	0.7432	I
5	0.4753	0.4753	I	0.5006	0.5006	I	0.5182	0.5182	I
6	0.5681	2.5320	C	0.6102	2.3259	C	0.6419	2.1522	C
7	0.8023	31.5784	C	0.8281	30.4194	C	0.8458	29.6339	C
8	0.2489	0.7153	I	0.3390	0.6809	I	0.4095	0.6581	I
9	1.0662	1.0662	D	1.0610	1.0610	D	1.0579	1.0579	D
10	1.0741	1.0741	D	1.0736	1.0736	D	1.0739	1.0739	D
11	1.0721	1.0721	D	1.0588	1.0588	D	1.0582	1.0582	D
12	1.2313	1.2993	D	1.2716	1.2716	D	1.2798	1.2798	D
13	0.9601	2.5489	C	1.0051	2.3755	D	1.0372	2.2506	D
14	1.2797	2.2669	D	1.3472	2.0266	D	1.4042	1.8454	D
15	0.9840	2.2035	C	1.0172	2.0512	D	1.0413	1.9482	D
16	0.8081	∞	C	0.8305	∞	C	0.8460	∞	C
17	0.8475	0.8475	I	0.9691	0.9691	I	1.0536	1.0536	D
18	0.9337	0.9337	I	0.9312	0.9312	I	0.9257	0.9257	I
19	0.6328	2.1413	C	0.6705	1.7111	C	0.6982	1.4534	C
20	1.2643	1.7236	D	1.3759	1.5772	D	1.4864	1.4864	D
21	1.1441	1.1441	D	1.1371	1.1371	D	1.1313	1.1313	D
22	1.0073	1.0073	D	1.0098	1.0098	D	1.0392	1.0392	D
23	1.3117	1.3117	D	1.4484	1.4484	D	1.3763	1.3763	D
24	0.0000	0.7253	I	0.0000	0.6612	I	0.0000	0.6152	I
25	1.2101	2.0733	D	1.2735	1.9498	D	1.3158	1.8777	D
26	0.0000	0.8308	I	0.0000	0.7425	I	0.0000	0.6903	I
27	0.9179	0.9179	I	0.9787	0.9787	I	1.0220	1.0220	D
28	0.7959	1.1565	C	0.9199	0.9443	I	0.9253	0.9253	I
29	0.0000	1.4812	C	0.0000	1.3102	C	0.0000	1.2039	C
30	0.6413	1.2982	C	0.6783	1.1251	C	0.7053	1.0264	C
31	0.8851	2.1450	C	0.9424	2.0216	C	0.9847	1.9430	C
32	0.8207	0.8207	I	0.8031	0.8031	I	0.7895	0.7895	I
33	1.1382	1.1382	D	1.1068	1.1068	D	1.0877	1.0877	D
34	0.9556	0.9556	I	0.9368	0.9368	I	0.7996	0.7996	I
35	0.8088	0.9547	I	0.9624	0.9624	I	0.9980	0.9980	I
36	1.0613	1.0613	D	1.0643	1.0643	D	1.0656	1.0656	D
37	0.0000	1.1623	C	0.0000	1.0666	C	0.0000	1.0049	C
38	1.1424	2.6807	D	1.1556	2.4569	D	1.1650	2.3160	D
39	0.3489	2.4647	C	0.3724	2.2332	C	0.3898	2.0899	C
40	1.1059	1.1059	D	1.1192	1.1192	D	1.1304	1.1304	D
41	1.1115	1.1115	D	1.0731	1.0731	D	1.0463	1.0463	D

(continued)

Table 6.6 (continued)

	$\gamma = 0.1$			$\gamma = 0.3$			$\gamma = 0.5$		
States	ε^-	ε^+	RTS	ε^-	ε^+	RTS	ε^-	ε^+	RTS
42	0.7221	0.7221	I	0.7461	0.7461	I	0.7598	0.7598	I
43	0.6211	1.9227	C	0.6739	1.8405	C	0.7108	1.7525	C
44	0.9155	2.2560	C	0.9675	2.0501	C	1.0072	1.9251	D
45	1.0363	1.0363	D	0.9742	0.9742	I	0.9342	0.9342	I
46	0.9501	0.9501	I	1.0011	1.0011	D	1.0471	1.0471	D
47	1.0504	1.0504	D	1.0858	1.0858	D	1.0972	1.0972	D
48	0.1195	1.1979	C	0.1280	1.0559	C	0.1333	0.9704	I

Note: I = IRS, C = CRS, and D = DRS

level that forms an ellipse around the mean value of the inputs and outputs of each firm increases or decreases. As a result, a firm might experience a change in the RTS status for a change in γ. The relevant question, therefore, before any state facing uncertainty is: at what level of γ should it consider making an appropriate judgment about its scale decision? Since our proposed stochastic approach generates only expected SE scores rather than distributions of random SE scores, we find it difficult to answer this question. Our focus in this chapter is, however, on the ex post facto analysis of already effected decisions, which can have some uses in the control aspects of management where evaluations of the RTS performance are required. What is left unattended here is the *ex-ante* (planning) problem of how to use this knowledge to arrive at the possible value of γ, which can be used while effecting future-oriented decisions regarding whether to scale up or scale down the operations.[3]

Furthermore, it can be noticed that for $\beta = 0.99$ (Table 6.6), the upper-bound scale elasticity of state 16 is unbounded. This means that any proportional increase in output lies outside the technology set. However, in Table 6.7, both the lower and upper bounds of the SEs of state 16 are indicated by a dashed line, implying that model (6.12) is infeasible because by changing β from 1 to 1.01, these units appear outside the feasible region. Therefore, program (6.12) becomes infeasible when there are no data available outside the feasible region to determine RTS (Grosskopf, 1996).

[3] One may also adopt the probabilistic approach by Kao and Liu (2009, 2019) to compute the distributions of the SE scores, which are more informative for a firm manager to arrive at an appropriate level of γ for its scale decision, which we have pointed out in our concluding remarks as a potential avenue for future research.

Table 6.7 The stochastic value-based scale elasticity scores of 48 states ($\beta = 1.01$)

	$\gamma = 0.1$			$\gamma = 0.3$			$\gamma = 0.5$		
States	ε^-	ε^+	RTS	ε^-	ε^+	RTS	ε^-	ε^+	RTS
1	1.0397	1.0397	D	1.0398	1.0398	D	1.0401	1.0401	D
2	0.5800	2.2292	C	0.5962	1.9174	C	0.6091	1.6914	C
3	1.1896	1.1896	D	1.0576	1.0576	D	1.0548	1.0548	D
4	0.7047	0.8353	I	0.7478	0.7478	I	0.7470	0.7470	I
5	0.4803	0.4803	I	0.5056	0.5056	I	0.5232	0.5232	I
6	0.6073	2.5975	C	0.6448	2.3381	C	0.6731	2.1631	C
7	0.9471	26.0339	C	0.9696	25.2789	C	0.9847	24.7588	C
8	0.3037	0.7353	I	0.3808	0.6964	I	0.4468	0.6706	I
9	1.2336	1.2336	D	1.0597	1.0597	D	1.0567	1.0567	D
10	1.1405	1.1405	D	1.1003	1.1003	D	1.0723	1.0723	D
11	1.0705	1.0705	D	1.0630	1.0630	D	1.0627	1.0627	D
12	1.2434	1.3215	D	1.2648	1.2648	D	1.2727	1.2727	D
13	0.9869	2.5890	C	1.0253	2.3890	D	1.0532	2.2613	D
14	1.3440	2.2604	D	1.3889	2.0173	D	1.4299	1.8693	D
15	1.0169	2.2914	D	1.0448	2.1194	D	1.0656	2.0042	D
16	–	–		–	–		–	–	
17	0.8501	0.8501	I	0.9697	0.9697	I	1.0525	1.0525	D
18	1.2970	1.2970	D	1.2035	1.2035	D	1.1418	1.1418	D
19	0.7084	2.3476	C	0.7145	1.9231	C	0.7209	1.6076	C
20	1.2950	1.7157	D	1.3851	1.5664	D	1.4723	1.4723	D
21	1.1778	1.1778	D	1.1340	1.1340	D	1.1284	1.1284	D
22	1.0385	1.0385	D	1.0383	1.0383	D	1.0384	1.0384	D
23	1.3791	1.3791	D	1.4373	1.4373	D	1.4970	1.4970	D
24	0.0000	0.7440	I	0.0000	0.6763	I	0.0000	0.6292	I
25	1.2960	2.0887	D	1.3374	1.9593	D	1.3658	1.8847	D
26	0.0000	0.8662	I	0.0000	0.7724	I	0.0000	0.7171	I
27	0.9194	0.9194	I	0.9791	0.9791	I	1.0216	1.0216	D
28	0.8246	1.1742	C	0.9251	0.9496	C	0.9266	0.9266	I
29	0.0000	1.5045	C	0.0000	1.3315	C	0.0000	1.2243	C
30	0.6637	1.3272	C	0.6950	1.1456	C	0.7184	1.0419	C
31	0.9043	2.1907	C	0.9559	2.0573	C	0.9960	1.9724	C
32	0.8236	0.8236	I	0.8062	0.8062	I	0.7928	0.7928	I
33	1.1350	1.1350	D	1.1056	1.1056	D	1.0858	1.0858	D
34	0.9565	0.9565	I	0.9450	0.9450	I	0.9331	0.9331	I
35	0.8183	0.9725	I	0.9631	0.9631	I	1.0122	1.0122	D
36	1.0601	1.0601	D	1.0630	1.0630	D	1.0643	1.0643	D
37	0.0000	1.1898	C	0.0000	1.0926	C	0.0000	1.0296	C
38	1.2669	2.6820	D	1.2550	2.4582	D	1.2528	2.3170	D
39	0.3808	2.5053	C	0.4017	2.2727	C	0.4174	2.1277	C
40	1.1099	1.1099	D	1.1166	1.1166	D	1.1275	1.1275	D
41	1.1090	1.1090	D	1.0715	1.0715	D	1.0453	1.0453	D

(continued)

Table 6.7 (continued)

	$\gamma = 0.1$			$\gamma = 0.3$			$\gamma = 0.5$		
States	ε^-	ε^+	RTS	ε^-	ε^+	RTS	ε^-	ε^+	RTS
42	0.7261	0.7261	I	0.7724	0.7724	I	0.8102	0.8102	I
43	0.6704	1.9775	C	0.7192	1.8697	C	0.7528	1.7736	C
44	0.9380	2.2850	C	0.9848	2.0729	C	1.0212	1.9434	D
45	1.0355	1.0355	D	0.9747	0.9747	I	0.9354	0.9354	I
46	0.9511	0.9511	I	1.0011	1.0011	D	1.0461	1.0461	D
47	1.0493	1.0493	D	1.0839	1.0839	D	1.0951	1.0951	D
48	0.1429	1.2178	C	0.1492	1.0739	C	0.1532	0.9872	I

Note: I = IRS, C = CRS, and D = DRS

6.3.4 Deterministic Value-Based SE Scores of the States

In the spirit of program (6.17), we ran the valued-based linear program (6.4) to compute the lower and upper bounds of the SE scores of 48 states in a deterministic environment for $\beta = 0.99$ and 1.01, which are presented in Table 6.8. The results show that the same 21 efficient states that are found fully technically efficient in a stochastic environment are also efficient here and the remaining 27 states are inefficient. As regards their RTS characterizations, for the case of $\beta = 1.01$, we find ten efficient states (*i.e.*, 2, 6, 7, 19, 29, 30, 31, 37, 39, and 43) exhibiting CRS. For four efficient states (*i.e.*, 8, 24, 26, and 48), the RTS are increasing since their upper bounds of the SE scores are all less than unity. And finally, six efficient states (*i.e.*, 13, 14, 15, 25, 38, and 44) are exhibiting DRS. For the remaining state 16, both the lower- and upper-bound SEs are indicated by dashed lines. Hence, as discussed in Sect. 6.3.3, there are no data available to determine the RTS type. This is because as we consider β from 1 to 1.01, these states appear outside the feasible region. Therefore, program (6.3) becomes infeasible when there are no data available outside the feasible region to determine their RTS possibilities.

For $\beta = 0.99$, the inefficient states 18 and 35 exhibit IRS; however, these same states exhibit DRS at $\beta = 1.01$. Moreover, while the RTS types of state 16 for the case of $\beta = 1.01$ were not determined, they are constant at $\beta = 0.99$.

6.3.5 A Comparison Between the Deterministic and Stochastic Methods

On the comparison between deterministic and stochastic TE estimates, a few observations are worth mentioning. First, both the deterministic and stochastic methods are in complete agreement in declaring the same 21 states as fully technically efficient. Second, both the methods yield the same value-based TE efficiency scores of states for $\gamma = 0.5$. This finding is as per our expectations (see Theorem 6.2), which has led us to conclude that the deterministic technology is a special case of stochastic

Table 6.8 Deterministic value-based TE and SE scores of the states

State	TE	$\beta = 1.01$			$\beta = 0.99$			State	TE	$\beta = 1.01$			$\beta = 0.99$		
		ε^-	ε^+		ε^-	ε^+				ε^-	ε^+		ε^-	ε^+	
1	0.868	1.0401	1.0401	D	1.0410	1.0410	D	25	1	1.3658	1.8847	D	1.3158	1.8777	D
2	1	0.6091	1.6914	C	0.5848	1.6184	C	26	1	0.0000	0.7171	I	0.0000	0.6903	I
3	0.948	1.0548	1.0548	D	1.0459	1.0459	D	27	0.942	1.0216	1.0216	D	1.0220	1.0220	D
4	0.995	0.7470	0.7470	I	0.7432	0.7432	I	28	0.991	0.9266	0.9266	I	0.9253	0.9253	I
5	0.990	0.5232	0.5232	I	0.5182	0.5182	I	29	1	0.0000	1.2243	C	0.0000	1.2039	C
6	1	0.6731	2.1631	C	0.6419	2.1522	C	30	1	0.7184	1.0419	C	0.7053	1.0264	C
7	1	0.9847	24.7588	C	0.8458	29.6339	C	31	1	0.9960	1.9724	C	0.9847	1.9430	C
8	1	0.4468	0.6706	I	0.4095	0.6581	I	32	0.926	0.7928	0.7928	I	0.7895	0.7895	I
9	0.960	1.0567	1.0567	D	1.0579	1.0579	D	33	0.941	1.0858	1.0858	D	1.0877	1.0877	D
10	0.899	1.0723	1.0723	D	1.0739	1.0739	D	34	0.881	0.9331	0.9331	I	0.7996	0.7996	I
11	0.906	1.0627	1.0627	D	1.0582	1.0582	D	35	0.993	1.0122	1.0122	D	0.9980	0.9980	I
12	0.992	1.2727	1.2727	D	1.2798	1.2798	D	36	0.931	1.0643	1.0643	D	1.0656	1.0656	D
13	1	1.0532	2.2613	D	1.0372	2.2506	D	37	1	0.0000	1.0296	C	0.0000	1.0049	C
14	1	1.4299	1.8693	D	1.4042	1.8454	D	38	1	1.2528	2.3170	D	1.1650	2.3160	D
15	1	1.0656	2.0042	D	1.0413	1.9482	D	39	1	0.4174	2.1277	C	0.3898	2.0899	C
16	1	–	–		0.8460	∞	C	40	0.930	1.1275	1.1275	D	1.1304	1.1304	D
17	0.892	1.0525	1.0525	D	1.0536	1.0536	D	41	0.956	1.0453	1.0453	D	1.0463	1.0463	D
18	0.939	1.1418	1.1418	D	0.9257	0.9257	I	42	0.893	0.8102	0.8102	I	0.7598	0.7598	I
19	1	0.7209	1.6076	C	0.6982	1.4534	C	43	1	0.7528	1.7736	C	0.7108	1.7525	C
20	0.998	1.4723	1.4723	D	1.4864	1.4864	D	44	1	1.0212	1.9434	D	1.0072	1.9251	D
21	0.886	1.1284	1.1284	D	1.1313	1.1313	D	45	0.967	0.9354	0.9354	I	0.9342	0.9342	I
22	0.862	1.0384	1.0384	D	1.0392	1.0392	D	46	0.904	1.0461	1.0461	D	1.0471	1.0471	D
23	0.958	1.4970	1.4970	D	1.3763	1.3763	D	47	0.923	1.0951	1.0951	D	1.0972	1.0972	D
24	1	0.0000	0.6292	I	0.0000	0.6152	I	48	1	0.1532	0.9872	I	0.1333	0.9704	I

Note: I = IRS, C = CRS, and D = DRS

technology for $\gamma = 0.5$ in exhibiting the same TE behavior of states. Third, the stochastic TE scores of the inefficient states are higher compared to their deterministic counterparts. This finding is also as per our expectations because the very purpose of our chance-constrained formulation is to allow the observed inputs and output of states to cross the efficiency frontier, but not too often. For any given observed input–output vector, the efficiency frontier in the stochastic case is now located closer than before. Finally, the efficiency scores of the states increase with sharpening (decreasing) the tolerance level of the chance constraints.

As regards the comparison between the deterministic and stochastic SE estimates, we find some apparent differences in their RTS characterizations. First, although the RTS types of the remaining states are the same in both the deterministic and stochastic methods, the degree of RTS (*i.e.*, lower and upper bounds of SE) varies to some degree. For example, consider state 15. In the case of $\beta = 0.99$ and $\gamma = 0.1$, the stochastic lower and upper bounds of the SE estimates suggest that a 1% increase in input raises the output by 0.9840% and 2.2035%, respectively. However, for $\beta = 0.99$ in the deterministic case, the lower and upper SE bounds suggest that a 1% increase in inputs yields an increase in output of 1.0413% and 1.9482%, respectively. Second, for $\gamma = 0.5$, the value-based SE scores of states are found to be the same between the two methods. As per Theorem 6.2, this finding is not contrary to our expectations. One can, therefore, conclude that the deterministic technology is again a special case of the stochastic technology for $\gamma = 0.5$ in exhibiting the same RTS characterization of states.

References

Atici, K. B., & Podinovski, V. V. (2015). Using data envelopment analysis for the assessment of technical efficiency of units with different specializations: An application to agriculture. *Omega, 54*, 72–83.

Banker, R. D., Charnes, A., & Cooper, W. W. (1984). Some models for estimating technical and scale inefficiencies in data envelopment analysis. *Management Science, 30*, 1078–1092.

Banker, R. D., Cooper, W. W., Seiford, L. M., Thrall, R. M., & Zhu, J. (2004). Returns to scale in different DEA models. *European Journal of Operational Research, 154*, 345–362.

Camanho, A. S., & Dyson, R. G. (2005). Cost efficiency measurement with price uncertainty: A DEA application to bank branch assessments. *European Journal of Operational Research, 161*, 432–446.

Charnes, A., & Cooper, W. W. (1977). Goal programming and multiple objective optimizations: Part 1. *European Journal of Operational Research, 1*, 39–54.

Charnes, A., Cooper, W. W., & Rhodes, E. (1978). Measuring the efficiency of decision-making units. *European Journal of Operational Research, 2*, 429–444.

Cherchye, L., Rock, B. D., Dierynck, B., Roodhooft, F., & Sabbe, J. (2013). Opening the "black box" of efficiency measurement: Input allocation in multioutput settings. *Operations Research, 61*, 1148–1165.

Cherchye, L., De Rock, B., & Walheer, B. (2015). Multi-output efficiency with good and bad outputs. *European Journal of Operational Research, 240*, 872–881.

Cherchye, L., De Rock, B., & Walheer, B. (2016). Multi-output profit efficiency and directional distance functions. *Omega, 61*, 100–109.

Cook, W. D., & Zhu, J. (2011). Multiple variable proportionality in data envelopment analysis. *Operations Research, 59*, 1024–1032.

Cooper, W. W., Deng, H., Huang, Z., & Li, S. X. (2002). Chance constrained programming approaches to technical efficiencies and inefficiencies in stochastic data envelopment analysis. *Journal of the Operational Research Society, 53*, 1347–1356.

Cooper, W., Huang, Z., & Li, S. X. (1996). Satisficing DEA models under chance constraints. *Annals of Operations Research, 66*, 279–295.

Cooper, W. W., Seiford, L. M., & Tone, K. (2000). Data envelopment analysis. In W. W. Cooper, L. M. Seiford, & J. Zhu (Eds.), *Handbook on data envelopment analysis* (1st ed., pp. 1–40).

Färe, R., Grosskopf, S., & Lovell, C. A. K. (1985). *The measurement of efficiency of production*. Kluwer Nijhoff.

Farrell, M. J. (1957). The measurement of productive efficiency. *Journal of the Royal Statistical Society (Series A), 120*, 253–290.

Fukuyama, H., & Weber, W. L. (2008). Profit inefficiency of Japanese securities firms. *Journal of Applied Economics, 11*, 281–303.

Grosskopf, S. (1996). Statistical inference and nonparametric efficiency: A selective survey. *Journal of Productivity Analysis, 7*, 161–176.

Huang, Z., & Li, S. X. (1996). Dominance stochastic models in data envelopment analysis. *European Journal of Operational Research, 95*, 390–403.

Huang, Z., & Li, S. X. (2001). Stochastic DEA models with different types of input-output disturbances. *Journal of Productivity Analysis, 15*, 95–113.

Kahane, Y. (1977). Determination of the product mix and the business policy of an insurance company—a portfolio approach. *Management Science, 23*, 1060–1069.

Kao, C., & Liu, S. T. (2009). Stochastic data envelopment analysis in measuring the efficiency of Taiwan commercial banks. *European Journal of Operational Research, 196*, 312–322.

Kao, C., & Liu, S. T. (2019). Stochastic efficiency measures for production units with correlated data. *European Journal of Operational Research, 273*, 278–287.

Kuosmanen, T. (2005). Weak disposability in nonparametric production analysis with undesirable outputs. *American Journal of Agricultural Economics, 87*, 1077–1082.

Kuosmanen, T., & Podinovski, V. (2009). Weak disposability in nonparametric production analysis: Reply to Färe and Grosskopf. *American Journal of Agricultural Economics, 91*, 539–545.

Land, K. C., Lovell, C. A. K., & Thore, S. (1993). Chance constrained data envelopment analysis. *Managerial and Decision Economics, 14*, 541–554.

Li, S. X. (1995a). A satisficing chance constrained model in the portfolio selection of insurance lines and investments. *Journal of the Operational Research Society, 46*(9), 1111–1120.

Li, S. X. (1995b). An insurance and investment portfolio model using chance constrained programming. *Omega, 23*(5), 577–585.

Li, S. X. (1998). Stochastic models and variable returns to scales in data envelopment analysis. *European Journal of Operational Research, 104*, 532–548.

Li, S. X., & Huang, Z. (1996). Determination of the portfolio selection for a property-liability insurance company. *European Journal of Operational Research, 88*(2), 257–268.

Olesen, O. B., & Petersen, N. C. (2000). Foundation of chance constrained data envelopment analysis for pareto-Koopmann efficient production possibility sets. In *International DEA Symposium 2000, Measurement and Improvement in the 21st Century* (pp. 313–349). The University of Queensland.

Podinovski, V. V. (2004a). Production trade-offs and weight restrictions in data envelopment analysis. *Journal of the Operational Research Society, 55*, 1311–1322.

Podinovski, V. V. (2004b). Bridging the gap between the constant and variable returns-to-scale models: Selective proportionality in data envelopment analysis. *Journal of the Operational Research Society, 55*, 265–276.

Podinovski, V. V. (2007). Improving data envelopment analysis by the use of production trade-offs. *Journal of the Operational Research Society, 58*, 1261–1270.

Podinovski, V. V. (2015). DEA models with production trade-offs and weight restrictions. In W. D. Cook & J. Zhu (Eds.), *Data envelopment analysis* (pp. 105–144). Springer.

Podinovski, V. V. (2016). Optimal weights in DEA models with weight restrictions. *European Journal of Operational Research, 254*, 916–924.

Ray, S. C., Chen, L., & Mukherjee, L. (2008). Input price variation across locations and a generalized measure of cost-efficiency. *International Journal of Production Economics, 116*, 208–218.

Sahoo, B. K., & Tone, K. (2013). Non-parametric measurement of economies of scale and scope in non-competitive environment with price uncertainty. *Omega, 41*, 97–111.

Sahoo, B. K., Zhu, J., & Tone, K. (2014a). Decomposing efficiency and returns to scale in two-stage network systems. In W. D. Cook & J. Zhu (Eds.), *Data envelopment analysis* (pp. 137–164). Springer.

Sahoo, B. K., Zhu, J., Tone, K., & Klemen, B. M. (2014b). Decomposing technical efficiency and scale elasticity in two-stage network DEA. *European Journal of Operational Research, 233*, 584–594.

Sengupta, J. K. (2000). *Dynamic and stochastic efficiency analysis: Economics of data envelopment analysis*. World Scientific.

Sharpe, W. F. (1963). A simplified model for portfolio analysis. *Management Science, 9*, 277–293.

Sueyoshi, T. (1997). Measuring efficiencies and returns to scale of Nippon telegraph & telephone in production and cost analyses. *Management Science, 43*, 779–796.

Sueyoshi, T., & Goto, N. (2012). Returns to scale and damages to scale under natural and manageriam disposability: Strategy, efficiency and competitiveness of petroleum firms. *Energy Economics, 34*, 645–662.

Tone, K. (2001). On returns to scale under weight restrictions in data envelopment analysis. *Journal of Productivity Analysis, 16*, 31–47.

Tone, K., & Sahoo, B. K. (2005). Evaluating cost efficiency and returns to scale in the Life Insurance Corporation of India using data envelopment analysis. *Socio-Economic Planning Sciences, 39*, 261–285.

Tone, K., & Sahoo, B. K. (2006). Re-examining scale elasticity in DEA. *Annals of Operations Research, 145*, 69–87.

Printed by Printforce, the Netherlands